Anonymous

The Elements of Hydraulic Engineering

Vol. 4

Anonymous

The Elements of Hydraulic Engineering
Vol. 4

ISBN/EAN: 9783337214517

Printed in Europe, USA, Canada, Australia, Japan

Cover: Foto ©berggeist007 / pixelio.de

More available books at **www.hansebooks.com**

THE ELEMENTS

OF

HYDRAULIC ENGINEERING

Prepared for Students of
The International Correspondence Schools
SCRANTON, PA.

Volume IV

TABLES AND FORMULAS

First Edition

SCRANTON
THE COLLIERY ENGINEER CO.
1898

BURR PRINTING HOUSE,
FRANKFORT AND JACOB STREETS,
NEW YORK.

TABLES AND FORMULAS.

This volume contains all the principal Tables and Formulas which are likely to be used by the student in practice. They have been collected and placed in this volume in order to make them convenient for ready reference, so that the student will not be obliged to hunt them out in the preceding volumes. The number after each formula is the same as the number following the same formula in the text.

TABLE

OF

COMMON LOGARITHMS

OF NUMBERS

From 1 to 10,000.

N.	Log.	N.	Log.	N.	Log.	N.	Log.	N.	Log.
0	— ∞	20	30 103	40	60 206	60	77 815	80	90 309
1	00 000	21	32 222	41	61 278	61	78 533	81	90 849
2	30 103	22	34 242	42	62 325	62	79 239	82	91 381
3	47 712	23	36 173	43	63 347	63	79 934	83	91 908
4	60 206	24	38 021	44	64 345	64	80 618	84	92 428
5	69 897	25	39 794	45	65 321	65	81 291	85	92 942
6	77 815	26	41 497	46	66 276	66	81 954	86	93 450
7	84 510	27	43 136	47	67 210	67	82 607	87	93 952
8	90 309	28	44 716	48	68 124	68	83 251	88	94 448
9	95 424	29	46 240	49	69 020	69	83 885	89	94 939
10	00 000	30	47 712	50	69 897	70	84 510	90	95 424
11	04 139	31	49 136	51	70 757	71	85 126	91	95 904
12	07 918	32	50 515	52	71 600	72	85 733	92	96 379
13	11 394	33	51 851	53	72 428	73	86 332	93	96 848
14	14 613	34	53 148	54	73 239	74	86 923	94	97 313
15	17 609	35	54 407	55	74 036	75	87 506	95	97 772
16	20 412	36	55 630	56	74 819	76	88 081	96	98 227
17	23 045	37	56 820	57	75 587	77	88 649	97	98 677
18	25 527	38	57 978	58	76 343	78	89 209	98	99 123
19	27 875	39	59 106	59	77 085	79	89 763	99	99 564
20	30 103	40	60 206	60	77 815	80	90 309	100	00 000

N.	L. 0	1	2	3	4	5	6	7	8	9
100	00 000	043	087	130	173	217	260	303	346	389
101	432	475	518	561	604	647	689	732	775	817
102	860	903	945	988	*030	*072	*115	*157	*199	*242
103	01 284	326	368	410	452	494	536	578	620	662
104	703	745	787	828	870	912	953	995	*036	*078
105	02 119	160	202	243	284	325	366	407	449	490
106	531	572	612	653	694	735	776	816	857	898
107	938	979	*019	*060	*100	*141	*181	*222	*262	*302
108	03 342	383	423	463	503	543	583	623	663	703
109	743	782	822	862	902	941	981	*021	*060	*100
110	04 139	179	218	258	297	336	376	415	454	493
111	532	571	610	650	689	727	766	805	844	883
112	922	961	999	*038	*077	*115	*154	*192	*231	*269
113	05 308	346	385	423	461	500	538	576	614	652
114	690	729	767	805	843	881	918	956	994	*032
115	06 070	108	145	183	221	258	296	333	371	408
116	446	483	521	558	595	633	670	707	744	781
117	819	856	893	930	967	*004	*041	*078	*115	*151
118	07 188	225	262	298	335	372	408	445	482	518
119	555	591	628	664	700	737	773	809	846	882
120	918	954	990	*027	*063	*099	*135	*171	*207	*243
121	08 279	314	350	386	422	458	493	529	565	600
122	636	672	707	743	778	814	849	884	920	955
123	991	*026	*061	*096	*132	*167	*202	*237	*272	*307
124	09 342	377	412	447	482	517	552	587	621	656
125	691	726	760	795	830	864	899	934	968	*003
126	10 037	072	106	140	175	209	243	278	312	346
127	380	415	449	483	517	551	585	619	653	687
128	721	755	789	823	857	890	924	958	992	*025
129	11 059	093	126	160	193	227	261	294	327	361
130	394	428	461	494	528	561	594	628	661	694
131	727	760	793	826	860	893	926	959	992	*024
132	12 057	090	123	156	189	222	254	287	320	352
133	385	418	450	483	516	548	581	613	646	678
134	710	743	775	808	840	872	905	937	969	*001
135	13 033	066	098	130	162	194	226	258	290	322
136	354	386	418	450	481	513	545	577	609	640
137	672	704	735	767	799	830	862	893	925	956
138	988	*019	*051	*082	*114	*145	*176	*208	*239	*270
139	14 301	333	364	395	426	457	489	520	551	582
140	613	644	675	706	737	768	799	829	860	891
141	922	953	983	*014	*045	*076	*106	*137	*168	*198
142	15 229	259	290	320	351	381	412	442	473	503
143	534	564	594	625	655	685	715	746	776	806
144	836	866	897	927	957	987	*017	*047	*077	*107
145	16 137	167	197	227	256	286	316	346	376	406
146	435	465	495	524	554	584	613	643	673	702
147	732	761	791	820	850	879	909	938	967	997
148	17 026	056	085	114	143	173	202	231	260	289
149	319	348	377	406	435	464	493	522	551	580
150	609	638	667	696	725	754	782	811	840	869
N.	L. 0	1	2	3	4	5	6	7	8	9

P. P.

	44	43	42
1	4.4	4.3	4.2
2	8.8	8.6	8.4
3	13.2	12.9	12.6
4	17.6	17.2	16.8
5	22.0	21.5	21.0
6	26.4	25.8	25.2
7	30.8	30.1	29.4
8	35.2	34.4	33.6
9	39.6	38.7	37.8

	41	40	39
1	4.1	4.0	3.9
2	8.2	8.0	7.8
3	12.3	12.0	11.7
4	16.4	16.0	15.6
5	20.5	20.0	19.5
6	24.6	24.0	23.4
7	28.7	28.0	27.3
8	32.8	32.0	31.2
9	36.9	36.0	35.1

	38	37	36
1	3.8	3.7	3.6
2	7.6	7.4	7.2
3	11.4	11.1	10.8
4	15.2	14.8	14.4
5	19.0	18.5	18.0
6	22.8	22.2	21.6
7	26.6	25.9	25.2
8	30.4	29.6	28.8
9	34.2	33.3	32.4

	35	34	33
1	3.5	3.4	3.3
2	7.0	6.8	6.6
3	10.5	10.2	9.9
4	14.0	13.6	13.2
5	17.5	17.0	16.5
6	21.0	20.4	19.8
7	24.5	23.8	23.1
8	28.0	27.2	26.4
9	31.5	30.6	29.7

	32	31	30
1	3.2	3.1	3.0
2	6.4	6.2	6.0
3	9.6	9.3	9.0
4	12.8	12.4	12.0
5	16.0	15.5	15.0
6	19.2	18.6	18.0
7	22.4	21.7	21.0
8	25.6	24.8	24.0
9	28.8	27.9	27.0

P. P.

N.	L. 0	1	2	3	4	5	6	7	8	9
150	17 609	638	667	696	725	754	782	811	840	869
151	898	926	955	984	*013	*041	*070	*099	*127	*156
152	18 184	213	241	270	298	327	355	384	412	441
153	469	498	526	554	583	611	639	667	696	724
154	752	780	808	837	865	893	921	949	977	*005
155	19 033	061	089	117	145	173	201	229	257	285
156	312	340	368	396	424	451	479	507	535	562
157	590	618	645	673	700	728	756	783	811	838
158	866	893	921	948	976	*003	*030	*058	*085	*112
159	20 140	167	194	222	249	276	303	330	358	385
160	412	439	466	493	520	548	575	602	629	656
161	683	710	737	763	790	817	844	871	898	925
162	952	978	*005	*032	*059	*085	*112	*139	*165	*192
163	21 219	245	272	299	325	352	378	405	431	458
164	484	511	537	564	590	617	643	669	696	722
165	748	775	801	827	854	880	906	932	958	985
166	22 011	037	063	089	115	141	167	194	220	246
167	272	298	324	350	376	401	427	453	479	505
168	531	557	583	608	634	660	686	712	737	763
169	789	814	840	866	891	917	943	968	994	*019
170	23 045	070	096	121	147	172	198	223	249	274
171	300	325	350	376	401	426	452	477	502	528
172	553	578	603	629	654	679	704	729	754	779
173	805	830	855	880	905	930	955	980	*005	*030
174	24 055	080	105	130	155	180	204	229	254	279
175	304	329	353	378	403	428	452	477	502	527
176	551	576	601	625	650	674	699	724	748	773
177	797	822	846	871	895	920	944	969	993	*018
178	25 042	066	091	115	139	164	188	212	237	261
179	285	310	334	358	382	406	431	455	479	503
180	527	551	575	600	624	648	672	696	720	744
181	768	792	816	840	864	888	912	935	959	983
182	26 007	031	055	079	102	126	150	174	198	221
183	245	269	293	316	340	364	387	411	435	458
184	482	505	529	553	576	600	623	647	670	694
185	717	741	764	788	811	834	858	881	905	928
186	951	975	998	*021	*045	*068	*091	*114	*138	*161
187	27 184	207	231	254	277	300	323	346	370	393
188	416	439	462	485	508	531	554	577	600	623
189	646	669	692	715	738	761	784	807	830	852
190	875	898	921	944	967	989	*012	*035	*058	*081
191	28 103	126	149	171	194	217	240	262	285	307
192	330	353	375	398	421	443	466	488	511	533
193	556	578	601	623	646	668	691	713	735	758
194	780	803	825	847	870	892	914	937	959	981
195	29 003	026	048	070	092	115	137	159	181	203
196	226	248	270	292	314	336	358	380	403	425
197	447	469	491	513	535	557	579	601	623	645
198	667	688	710	732	754	776	798	820	842	863
199	885	907	929	951	973	994	*016	*038	*060	*081
200	30 103	125	146	168	190	211	233	255	276	298
N.	L. 0	1	2	3	4	5	6	7	8	9

P. P.

	29	28
1	2.9	2.8
2	5.8	5.6
3	8.7	8.4
4	11.6	11.2
5	14.5	14.0
6	17.4	16.8
7	20.3	19.6
8	23.2	22.4
9	26.1	25.2

	27	26
1	2.7	2.6
2	5.4	5.2
3	8.1	7.8
4	10.8	10.4
5	13.5	13.0
6	16.2	15.6
7	18.9	18.2
8	21.6	20.8
9	24.3	23.4

	25
1	2.5
2	5.0
3	7.5
4	10.0
5	12.5
6	15.0
7	17.5
8	20.0
9	22.5

	24	23
1	2.4	2.3
2	4.8	4.6
3	7.2	6.9
4	9.6	9.2
5	12.0	11.5
6	14.4	13.8
7	16.8	16.1
8	19.2	18.4
9	21.6	20.7

	22	21
1	2.2	2.1
2	4.4	4.2
3	6.6	6.3
4	8.8	8.4
5	11.0	10.5
6	13.2	12.6
7	15.4	14.7
8	17.6	16.8
9	19.8	18.9

P. P.

LOGARITHMS.

N.	L. 0	1	2	3	4	5	6	7	8	9
200	30 103	125	146	168	190	211	233	255	276	298
201	320	341	363	384	406	428	449	471	492	514
202	535	557	578	600	621	643	664	685	707	728
203	750	771	792	814	835	856	878	899	920	942
204	963	984	*006	*027	*048	*069	*091	*112	*133	*154
205	31 175	197	218	239	260	281	302	323	345	366
206	387	408	429	450	471	492	513	534	555	576
207	597	618	639	660	681	702	723	744	765	785
208	806	827	848	869	890	911	931	952	973	994
209	32 015	035	056	077	098	118	139	160	181	201
210	222	243	263	284	305	325	346	366	387	408
211	428	449	469	490	510	531	552	572	593	613
212	634	654	675	695	715	736	756	777	797	818
213	838	858	879	899	919	940	960	980	*001	*021
214	33 041	062	082	102	122	143	163	183	203	224
215	244	264	284	304	325	345	365	385	405	425
216	445	465	486	506	526	546	566	586	606	626
217	646	666	686	706	726	746	766	786	806	826
218	846	866	885	905	925	945	965	985	*005	*025
219	34 044	064	084	104	124	143	163	183	203	223
220	242	262	282	301	321	341	361	380	400	420
221	439	459	479	498	518	537	557	577	596	616
222	635	655	674	694	713	733	753	772	792	811
223	830	850	869	889	908	928	947	967	986	*005
224	35 025	044	064	083	102	122	141	160	180	199
225	218	238	257	276	295	315	334	353	372	392
226	411	430	449	468	488	507	526	545	564	583
227	603	622	641	660	679	698	717	736	755	774
228	793	813	832	851	870	889	908	927	946	965
229	984	*003	*021	*040	*059	*078	*097	*116	*135	*154
230	36 173	192	211	229	248	267	286	305	324	342
231	361	380	399	418	436	455	474	493	511	530
232	549	568	586	605	624	642	661	680	698	717
233	736	754	773	791	810	829	847	866	884	903
234	922	940	959	977	996	*014	*033	*051	*070	*088
235	37 107	125	144	162	181	199	218	236	254	273
236	291	310	328	346	365	383	401	420	438	457
237	475	493	511	530	548	566	585	603	621	639
238	658	676	694	712	731	749	767	785	803	822
239	840	858	876	894	912	931	949	967	985	*003
240	38 021	039	057	075	093	112	130	148	166	184
241	202	220	238	256	274	292	310	328	346	364
242	382	399	417	435	453	471	489	507	525	543
243	561	578	596	614	632	650	668	686	703	721
244	739	757	775	792	810	828	846	863	881	899
245	917	934	952	970	987	*005	*023	*041	*058	*076
246	39 094	111	129	146	164	182	199	217	235	252
247	270	287	305	322	340	358	375	393	410	428
248	445	463	480	498	515	533	550	568	585	602
249	620	637	655	672	690	707	724	742	759	777
250	794	811	829	846	863	881	898	915	933	950
N.	L. 0	1	2	3	4	5	6	7	8	9

P. P.

	22	21
1	2.2	2.1
2	4.4	4.2
3	6.6	6.3
4	8.8	8.4
5	11.0	10.5
6	13.2	12.6
7	15.4	14.7
8	17.6	16.8
9	19.8	18.9

	20
1	2.0
2	4.0
3	6.0
4	8.0
5	10.0
6	12.0
7	14.0
8	16.0
9	18.0

	19
1	1.9
2	3.8
3	5.7
4	7.6
5	9.5
6	11.4
7	13.3
8	15.2
9	17.1

	18
1	1.8
2	3.6
3	5.4
4	7.2
5	9.0
6	10.8
7	12.6
8	14.4
9	16.2

	17
1	1.7
2	3.4
3	5.1
4	6.8
5	8.5
6	10.2
7	11.9
8	13.6
9	15.3

N.	L. 0	1	2	3	4	5	6	7	8	9
250	39 794	811	829	846	863	881	898	915	933	950
251	967	985	*002	*019	*037	*054	*071	*088	*106	*123
252	40 140	157	175	192	209	226	243	261	278	295
253	312	329	346	364	381	398	415	432	449	466
254	483	500	518	535	552	569	586	603	620	637
255	· 654	671	688	705	722	739	756	773	790	807
256	824	841	858	875	892	909	926	943	960	976
257	993	*010	*027	*044	*061	*078	*095	*111	*128	*145
258	41 162	179	196	212	229	246	263	280	296	313
259	330	347	363	380	397	414	430	447	464	481
260	497	514	531	547	564	581	597	614	631	647
261	664	681	697	714	731	747	764	780	797	814
262	830	847	863	880	896	913	929	946	963	979
263	996	*012	*029	*045	*062	*078	*095	*111	*127	*144
264	42 160	177	193	210	226	243	259	275	292	308
265	325	341	357	374	390	406	423	439	455	472
266	488	504	521	537	553	570	586	602	619	635
267	651	667	684	700	716	732	749	765	781	797
268	813	830	846	862	878	894	911	927	943	959
269	975	991	*008	*024	*040	*056	*072	*088	*104	*120
270	43 136	152	169	185	201	217	233	249	265	281
271	297	313	329	345	361	377	393	409	425	441
272	457	473	489	505	521	537	553	569	584	600
273	616	632	648	664	680	696	712	727	743	759
274	775	791	807	823	838	854	870	886	902	917
275	933	949	965	981	996	*012	*028	*044	*059	*075
276	44 091	107	122	138	154	170	185	201	217	232
277	248	264	279	295	311	326	342	358	373	389
278	404	420	436	451	467	483	498	514	529	545
279	560	576	592	607	623	638	654	669	685	700
280	716	731	747	762	778	793	809	824	840	855
281	871	886	902	917	932	948	963	979	994	*010
282	45 025	040	056	071	086	102	117	133	148	163
283	179	194	209	225	240	255	271	286	301	317
284	332	347	362	378	393	408	423	439	454	469
285	484	500	515	530	545	561	576	591	606	621
286	637	652	667	682	697	712	728	743	758	773
287	788	803	818	834	849	864	879	894	909	924
288	939	954	969	984	*000	*015	*030	*045	*060	*075
289	46 090	105	120	135	150	165	180	195	210	225
290	240	255	270	285	300	315	330	345	359	374
291	389	404	419	434	449	464	479	494	509	523
292	538	553	568	583	598	613	627	642	657	672
293	687	702	716	731	746	761	776	790	805	820
294	835	850	864	879	894	909	923	938	953	967
295	982	997	*012	*026	*041	*056	*070	*085	*100	*114
296	47 129	144	159	173	188	202	217	232	246	261
297	276	290	305	319	334	349	363	378	392	407
298	422	436	451	465	480	494	509	524	538	553
299	567	582	596	611	625	640	654	669	683	698
300	712	727	741	756	770	784	799	813	828	842
N.	L. 0	1	2	3	4	5	6	7	8	9

P. P.

18
1	1.8
2	3.6
3	5.4
4	7.2
5	9.0
6	10.8
7	12.6
8	14.4
9	16.2

17
1	1.7
2	3.4
3	5.1
4	6.8
5	8.5
6	10.2
7	11.9
8	13.6
9	15.3

16
1	1.6
2	3.2
3	4.8
4	6.4
5	8.0
6	9.6
7	11.2
8	12.8
9	14.4

15
1	1.5
2	3.0
3	4.5
4	6.0
5	7.5
6	9.0
7	10.5
8	12.0
9	13.5

14
1	1.4
2	2.8
3	4.2
4	5.6
5	7.0
6	8.4
7	9.8
8	11.2
9	12.6

N.	L. 0	1	2	3	4	5	6	7	8	9
300	47 712	727	741	756	770	784	799	813	828	842
301	857	871	885	900	914	929	943	958	972	986
302	48 001	015	029	044	058	073	087	101	116	130
303	144	159	173	187	202	216	230	244	259	273
304	287	302	316	330	344	359	373	387	401	416
305	430	444	458	473	487	501	515	530	544	558
306	572	586	601	615	629	643	657	671	686	700
307	714	728	742	756	770	785	799	813	827	841
308	855	869	883	897	911	926	940	954	968	982
309	996	*010	*024	*038	*052	*066	*080	*094	*108	*122
310	49 136	150	164	178	192	206	220	234	248	262
311	276	290	304	318	332	346	360	374	388	402
312	415	429	443	457	471	485	499	513	527	541
313	554	568	582	596	610	624	638	651	665	679
314	693	707	721	734	748	762	776	790	803	817
315	831	845	859	872	886	900	914	927	941	955
316	969	982	996	*010	*024	*037	*051	*065	*079	*092
317	50 106	120	133	147	161	174	188	202	215	229
318	243	256	270	284	297	311	325	338	352	365
319	379	393	406	420	433	447	461	474	488	501
320	515	529	542	556	569	583	596	610	623	637
321	651	664	678	691	705	718	732	745	759	772
322	786	799	813	826	840	853	866	880	893	907
323	920	934	947	961	974	987	*001	*014	*028	*041
324	51 055	068	081	095	108	121	135	148	162	175
325	188	202	215	228	242	255	268	282	295	308
326	322	335	348	362	375	388	402	415	428	441
327	455	468	481	495	508	521	534	548	561	574
328	587	601	614	627	640	654	667	680	693	706
329	720	733	746	759	772	786	799	812	825	838
330	851	865	878	891	904	917	930	943	957	970
331	983	996	*009	*022	*035	*048	*061	*075	*088	*101
332	52 114	127	140	153	166	179	192	205	218	231
333	244	257	270	284	297	310	323	336	349	362
334	375	388	401	414	427	440	453	466	479	492
335	504	517	530	543	556	569	582	595	608	621
336	634	647	660	673	686	699	711	724	737	750
337	763	776	789	802	815	827	840	853	866	879
338	892	905	917	930	943	956	969	982	994	*007
339	53 020	033	046	058	071	084	097	110	122	135
340	148	161	173	186	199	212	224	237	250	263
341	275	288	301	314	326	339	352	364	377	390
342	403	415	428	441	453	466	479	491	504	517
343	529	542	555	567	580	593	605	618	631	643
344	656	668	681	694	706	719	732	744	757	769
345	782	794	807	820	832	845	857	870	882	895
346	908	920	933	945	958	970	983	995	*008	*020
347	54 033	045	058	070	083	095	108	120	133	145
348	158	170	183	195	208	220	233	245	258	270
349	283	295	307	320	332	345	357	370	382	394
350	407	419	432	444	456	469	481	494	506	518
N.	L. 0	1	2	3	4	5	6	7	8	9

P. P.

15

1	1.5
2	3.0
3	4.5
4	6.0
5	7.5
6	9.0
7	10.5
8	12.0
9	13.5

14

1	1.4
2	2.8
3	4.2
4	5.6
5	7.0
6	8.4
7	9.8
8	11.2
9	12.6

13

1	1.3
2	2.6
3	3.9
4	5.2
5	6.5
6	7.8
7	9.1
8	10.4
9	11.7

12

1	1.2
2	2.4
3	3.6
4	4.8
5	6.0
6	7.2
7	8.4
8	9.6
9	10.8

N.	L. 0	1	2	3	4	5	6	7	8	9
350	54 407	419	432	444	456	469	481	494	506	518
351	531	543	555	568	580	593	605	617	630	642
352	654	667	679	691	704	716	728	741	753	765
353	777	790	802	814	827	839	851	864	876	888
354	900	913	925	937	949	962	974	986	998	*011
355	55 023	035	047	060	072	084	096	108	121	133
356	145	157	169	182	194	206	218	230	242	255
357	267	279	291	303	315	328	340	352	364	376
358	388	400	413	425	437	449	461	473	485	497
359	509	522	534	546	558	570	582	594	606	618
360	630	642	654	666	678	691	703	715	727	739
361	751	763	775	787	799	811	823	835	847	859
362	871	883	895	907	919	931	943	955	967	979
363	991	*003	*015	*027	*038	*050	*062	*074	*086	*098
364	56 110	122	134	146	158	170	182	194	205	217
365	229	241	253	265	277	289	301	312	324	336
366	348	360	372	384	396	407	419	431	443	455
367	467	478	490	502	514	526	538	549	561	573
368	585	597	608	620	632	644	656	667	679	691
369	703	714	726	738	750	761	773	785	797	808
370	820	832	844	855	867	879	891	902	914	926
371	937	949	961	972	984	996	*008	*019	*031	*043
372	57 054	066	078	089	101	113	124	136	148	159
373	171	183	194	206	217	229	241	252	264	276
374	287	299	310	322	334	345	.357	368	380	392
375	403	415	426	438	449	461	473	484	496	507
376	519	530	542	553	565	576	588	600	611	623
377	634	646	657	669	680	692	703	715	726	738
378	749	761	772	784	795	807	818	830	841	852
379	864	875	887	898	910	921	933	944	955	967
380	978	990	*001	*013	*024	*035	*047	*058	*070	*081
381	58 092	104	115	127	138	149	161	172	184	195
382	206	218	229	240	252	263	274	286	297	309
383	320	331	343	354	365	377	388	399	410	422
384	433	444	456	467	478	490	501	512	524	535
385	546	557	569	580	591	602	614	625	636	647
386	659	670	681	692	704	715	726	737	749	760
387	771	782	794	805	816	827	838	850	861	872
388	883	894	906	917	928	939	950	961	973	984
389	995	*006	*017	*028	*040	*051	*062	*073	*084	*095
390	59 106	118	129	140	151	162	173	184	195	207
391	218	229	240	251	262	273	284	295	306	318
392	329	340	351	362	373	384	395	406	417	428
393	439	450	461	472	483	494	506	517	528	539
394	550	561	572	583	594	605	616	627	638	649
395	660	671	682	693	704	715	726	737	748	759
396	770	780	791	802	813	824	835	846	857	868
397	879	890	901	912	923	934	945	956	966	977
398	988	999	*010	*021	*032	*043	*054	*065	*076	*086
399	60 097	108	119	130	141	152	163	173	184	195
400	206	217	228	239	249	260	271	282	293	304
N.	L. 0	1	2	3	4	5	6	7	8	9

P. P.

13
1	1.3
2	2.6
3	3.9
4	5.2
5	6.5
6	7.8
7	9.1
8	10.4
9	11.7

12
1	1.2
2	2.4
3	3.6
4	4.8
5	6.0
6	7.2
7	8.4
8	9.6
9	10.8

11
1	1.1
2	2.2
3	3.3
4	4.4
5	5.5
6	6.6
7	7.7
8	8.8
9	9.9

10
1	1.0
2	2.0
3	3.0
4	4.0
5	5.0
6	6.0
7	7.0
8	8.0
9	9.0

P. P.

N.	L. 0	1	2	3	4	5	6	7	8	9
400	60 206	217	228	239	249	260	271	282	293	304
401	314	325	336	347	358	369	379	390	401	412
402	423	433	444	455	466	477	487	498	509	520
403	531	541	552	563	574	584	595	606	617	627
404	638	649	660	670	681	692	703	713	724	735
405	746	756	767	778	788	799	810	821	831	842
406	853	863	874	885	895	906	917	927	938	949
407	959	970	981	991	*002	*013	*023	*034	*045	*055
408	61 066	077	087	098	109	119	130	140	151	162
409	172	183	194	204	215	225	236	247	257	268
410	278	289	300	310	321	331	342	352	363	374
411	384	395	405	416	426	437	448	458	469	479
412	490	500	511	521	532	542	553	563	574	584
413	595	606	616	627	637	648	658	669	679	690
414	700	711	721	731	742	752	763	773	784	794
415	805	815	826	836	847	857	868	878	888	899
416	909	920	930	941	951	962	972	982	993	*003
417	62 014	024	034	045	055	066	076	086	097	107
418	118	128	138	149	159	170	180	190	201	211
419	221	232	242	252	263	273	284	294	304	315
420	325	335	346	356	366	377	387	397	408	418
421	428	439	449	459	469	480	490	500	511	521
422	531	542	552	562	572	583	593	603	613	624
423	634	644	655	665	675	685	696	706	716	726
424	737	747	757	767	778	788	798	808	818	829
425	839	849	859	870	880	890	900	910	921	931
426	941	951	961	972	982	992	*002	*012	*022	*033
427	63 043	053	063	073	083	094	104	114	124	134
428	144	155	165	175	185	195	205	215	225	236
429	246	256	266	276	286	296	306	317	327	337
430	347	357	367	377	387	397	407	417	428	438
431	448	458	468	478	488	498	508	518	528	538
432	548	558	568	579	589	599	609	619	629	639
433	649	659	669	679	689	699	709	719	729	739
434	749	759	769	779	789	799	809	819	829	839
435	849	859	869	879	889	899	909	919	929	939
436	949	959	969	979	988	998	*008	*018	*028	*038
437	64 048	058	068	078	088	098	108	118	128	137
438	147	157	167	177	187	197	207	217	227	237
439	246	256	266	276	286	296	306	316	326	335
440	345	355	365	375	385	395	404	414	424	434
441	444	454	464	473	483	493	503	513	523	532
442	542	552	562	572	582	591	601	611	621	631
443	640	650	660	670	680	689	699	709	719	729
444	738	748	758	768	777	787	797	807	816	826
445	836	846	856	865	875	885	895	904	914	924
446	933	943	953	963	972	982	992	*002	*011	*021
447	65 031	040	050	060	070	079	089	099	108	118
448	128	137	147	157	167	176	186	196	205	215
449	225	234	244	254	263	273	283	292	302	312
450	321	331	341	350	360	369	379	389	398	408
N.	L. 0	1	2	3	4	5	6	7	8	9

P. P.

11
1 | 1.1
2 | 2.2
3 | 3.3
4 | 4.4
5 | 5.5
6 | 6.6
7 | 7.7
8 | 8.8
9 | 9.9

10
1 | 1.0
2 | 2.0
3 | 3.0
4 | 4.0
5 | 5.0
6 | 6.0
7 | 7.0
8 | 8.0
9 | 9.0

9
1 | 0.9
2 | 1.8
3 | 2.7
4 | 3.6
5 | 4.5
6 | 5.4
7 | 6.3
8 | 7.2
9 | 8.1

N.	L. 0	1	2	3	4	5	6	7	8	9
450	65 321	331	341	350	360	369	379	389	398	408
451	418	427	437	447	456	466	475	485	495	504
452	514	523	533	543	552	562	571	581	591	600
453	610	619	629	639	648	658	667	677	686	696
454	706	715	725	734	744	753	763	772	782	792
455	801	811	820	830	839	849	858	868	877	887
456	896	*906	916	925	935	944	954	963	973	982
457	992	*001	*011	*020	*030	*039	*049	*058	*068	*077
458	66 087	096	106	115	124	134	143	153	162	172
459	181	191	200	210	219	229	238	247	257	266
460	276	285	295	304	314	323	332	342	351	361
461	370	380	389	398	408	417	427	436	445	455
462	464	474	483	492	502	511	521	530	539	549
463	558	567	577	586	596	605	614	624	633	642
464	652	661	671	680	689	699	708	717	727	736
465	745	755	764	773	783	792	801	811	820	829
466	839	848	857	867	876	885	894	904	913	922
467	932	941	950	960	969	978	987	997	*006	*015
468	67 025	034	043	052	062	071	080	089	099	108
469	117	127	136	145	154	164	173	182	191	201
470	210	219	228	237	247	256	265	274	284	293
471	302	311	321	330	339	348	357	367	376	385
472	394	403	413	422	431	440	449	459	468	477
473	486	495	504	514	523	532	541	550	560	569
474	578	587	596	605	614	624	633	642	651	660
475	669	679	688	697	706	715	724	733	742	752
476	761	770	779	788	797	806	815	825	834	843
477	852	861	870	879	888	897	906	916	925	934
478	943	952	961	970	979	988	997	*006	*015	*024
479	68 034	043	052	061	070	079	088	097	106	115
480	124	133	142	151	160	169	178	187	196	205
481	215	224	233	242	251	260	269	278	287	296
482	305	314	323	332	341	350	359	368	377	386
483	395	404	413	422	431	440	449	458	467	476
484	485	494	502	511	520	529	538	547	556	565
485	574	583	592	601	610	619	628	637	646	655
486	664	673	681	690	699	708	717	726	735	744
487	753	762	771	780	789	797	806	815	824	833
488	842	851	860	869	878	886	895	904	913	922
489	931	940	949	958	966	975	984	993	*002	*011
490	69 020	028	037	046	055	064	073	082	090	099
491	108	117	126	135	144	152	161	170	179	188
492	197	205	214	223	232	241	249	258	267	276
493	285	294	302	311	320	329	338	346	355	364
494	373	381	390	399	408	417	425	434	443	452
495	461	469	478	487	496	504	513	522	531	539
496	548	557	566	574	583	592	601	609	618	627
497	636	644	653	662	671	679	688	697	705	714
498	723	732	740	749	758	767	775	784	793	801
499	810	819	827	836	845	854	862	871	880	888
500	897	906	914	923	932	940	949	958	966	975
N.	L. 0	1	2	3	4	5	6	7	8	9

P. P.

10
1	1.0
2	2.0
3	3.0
4	4.0
5	5.0
6	6.0
7	7.0
8	8.0
9	9.0

9
1	0.9
2	1.8
3	2.7
4	3.6
5	4.5
6	5.4
7	6.3
8	7.2
9	8.1

8
1	0.8
2	1.6
3	2.4
4	3.2
5	4.0
6	4.8
7	5.6
8	6.4
9	7.2

N.	L. 0	1	2	3	4	5	6	7	8	9	P. P.
500	69 897	906	914	923	932	940	949	958	966	975	
501	984	992	*001	*010	*018	*027	*036	*044	*053	*062	
502	70 070	079	088	096	105	114	122	131	140	148	
503	157	165	174	183	191	200	209	217	226	234	
504	243	252	260	269	278	286	295	303	312	321	
505	329	338	346	355	364	372	381	389	398	406	
506	415	424	432	441	449	458	467	475	484	492	
507	501	509	518	526	535	544	552	561	569	578	
508	586	595	603	612	621	629	638	646	655	663	
509	672	680	689	697	706	714	723	731	740	749	
510	757	766	774	783	791	800	808	817	825	834	
511	842	851	859	868	876	885	893	902	910	919	
512	927	935	944	952	961	969	978	986	995	*003	
513	71 012	020	029	037	046	054	063	071	079	088	
514	096	105	113	122	130	139	147	155	164	172	
515	181	189	198	206	214	223	231	240	248	257	
516	265	273	282	290	299	307	315	324	332	341	
517	349	357	366	374	383	391	399	408	416	425	
518	433	441	450	458	466	475	483	492	500	508	
519	517	525	533	542	550	559	567	575	584	592	
520	600	609	617	625	634	642	650	659	667	675	
521	684	692	700	709	717	725	734	742	750	759	
522	767	775	784	792	800	809	817	825	834	842	
523	850	858	867	875	883	892	900	908	917	925	
524	933	941	950	958	966	975	983	991	999	*008	
525	72 016	024	032	041	049	057	066	074	082	090	
526	099	107	115	123	132	140	148	156	165	173	
527	181	189	198	206	214	222	230	239	247	255	
528	263	272	280	288	296	304	313	321	329	337	
529	346	354	362	370	378	387	395	403	411	419	
530	428	436	444	452	460	469	477	485	493	501	
531	509	518	526	534	542	550	558	567	575	583	
532	591	599	607	616	624	632	640	648	656	665	
533	673	681	689	697	705	713	722	730	738	746	
534	754	762	770	779	787	795	803	811	819	827	
535	835	843	852	860	868	876	884	892	900	908	
536	916	925	933	941	949	957	965	973	981	989	
537	997	*006	*014	*022	*030	*038	*046	*054	*062	*070	
538	73 078	086	094	102	111	119	127	135	143	151	
539	159	167	175	183	191	199	207	215	223	231	
540	239	247	255	263	272	280	288	296	304	312	
541	320	328	336	344	352	360	368	376	384	392	
542	400	408	416	424	432	440	448	456	464	472	
543	480	488	496	504	512	520	528	536	544	552	
544	560	568	576	584	592	600	608	616	624	632	
545	640	648	656	664	672	679	687	695	703	711	
546	719	727	735	743	751	759	767	775	783	791	
547	799	807	815	823	830	838	846	854	862	870	
548	878	886	894	902	910	918	926	933	941	949	
549	957	965	973	981	989	997	*005	*013	*020	*028	
550	74 036	044	052	060	068	076	084	092	099	107	
N.	L. 0	1	2	3	4	5	6	7	8	9	P. P.

P. P.

9

1	0.9
2	1.8
3	2.7
4	3.6
5	4.5
6	5.4
7	6.3
8	7.2
9	8.1

8

1	0.8
2	1.6
3	2.4
4	3.2
5	4.0
6	4.8
7	5.6
8	6.4
9	7.2

7

1	0.7
2	1.4
3	2.1
4	2.8
5	3.5
6	4.2
7	4.9
8	5.6
9	6.3

N.	L. 0	1	2	3	4	5	6	7	8	9	P. P.
550	74 036	044	052	060	068	076	084	092	099	107	
551	115	123	131	139	147	155	162	170	178	186	
552	194	202	210	218	225	233	241	249	257	265	
553	273	280	288	296	304	312	320	327	335	343	
554	351	359	367	374	382	390	398	406	414	421	
555	429	437	445	453	461	468	476	484	492	500	
556	507	515	523	531	539	547	554	562	570	578	
557	586	593	601	609	617	624	632	640	648	656	
558	663	671	679	687	695	702	710	718	726	733	
559	741	749	757	764	772	780	788	796	803	811	
560	819	827	834	842	850	858	865	873	881	889	
561	896	904	912	920	927	935	943	950	958	966	
562	974	981	989	997	*005	*012	*020	*028	*035	*043	
563	75 051	059	066	074	082	089	097	105	113	120	
564	128	136	143	151	159	166	174	182	189	197	
565	205	213	220	228	236	243	251	259	266	274	
566	282	289	297	305	312	320	328	335	343	351	
567	358	366	374	381	389	397	404	412	420	427	
568	435	442	450	458	465	473	481	488	496	504	
569	511	519	526	534	542	549	557	565	572	580	
570	587	595	603	610	618	626	633	641	648	656	
571	664	671	679	686	694	702	709	717	724	732	
572	740	747	755	762	770	778	785	793	800	808	
573	815	823	831	838	846	853	861	868	876	884	
574	891	899	906	914	921	929	937	944	952	959	
575	967	974	982	989	997	*005	*012	*020	*027	*035	
576	76 042	050	057	065	072	080	087	095	103	110	
577	118	125	133	140	148	155	163	170	178	185	
578	193	200	208	215	223	230	238	245	253	260	
579	268	275	283	290	298	305	313	320	328	335	
580	343	350	358	365	373	380	388	395	403	410	
581	418	425	433	440	448	455	462	470	477	485	
582	492	500	507	515	522	530	537	545	552	559	
583	567	574	582	589	597	604	612	619	626	634	
584	641	649	656	664	671	678	686	693	701	708	
585	716	723	730	738	745	753	760	768	775	782	
586	790	797	805	812	819	827	834	842	849	856	
587	864	871	879	886	893	901	908	916	923	930	
588	938	945	953	960	967	975	982	989	997	*004	
589	77 012	019	026	034	041	048	056	063	070	078	
590	085	093	100	107	115	122	129	137	144	151	
591	159	166	173	181	188	195	203	210	217	225	
592	232	240	247	254	262	269	276	283	291	298	
593	305	313	320	327	335	342	349	357	364	371	
594	379	386	393	401	408	415	422	430	437	444	
595	452	459	466	474	481	488	495	503	510	517	
596	525	532	539	546	554	561	568	576	583	590	
597	597	605	612	619	627	634	641	648	656	663	
598	670	677	685	692	699	706	714	721	728	735	
599	743	750	757	764	772	779	786	793	801	808	
600	815	822	830	837	844	851	859	866	873	880	
N.	L. 0	1	2	3	4	5	6	7	8	9	P. P.

P. P.

8

1	0.8
2	1.6
3	2.4
4	3.2
5	4.0
6	4.8
7	5.6
8	6.4
9	7.2

7

1	0.7
2	1.4
3	2.1
4	2.8
5	3.5
6	4.2
7	4.9
8	5.6
9	6.3

N.	L. 0	1	2	3	4	5	6	7	8	9
600	77 815	822	830	837	844	851	859	866	873	880
601	887	895	902	909	916	924	931	938	945	952
602	960	967	974	981	988	996	*003	*010	*017	*025
603	78 032	039	046	053	061	068	075	082	089	097
604	104	111	118	125	132	140	147	154	161	168
605	176	183	190	197	204	211	219	226	233	240
606	247	254	262	269	276	283	290	297	305	312
607	319	326	333	340	347	355	362	369	376	383
608	390	398	405	412	419	426	433	440	447	455
609	462	469	476	483	490	497	504	512	519	526
610	533	540	547	554	561	569	576	583	590	597
611	604	611	618	625	633	640	647	654	661	668
612	675	682	689	696	704	711	718	725	732	739
613	746	753	760	767	774	781	789	796	803	810
614	817	824	831	838	845	852	859	866	873	880
615	888	895	902	909	916	923	930	937	944	951
616	958	965	972	979	986	993	*000	*007	*014	*021
617	79 029	036	043	050	057	064	071	078	085	092
618	099	106	113	120	127	134	141	148	155	162
619	169	176	183	190	197	204	211	218	225	232
620	239	246	253	260	267	274	281	288	295	302
621	309	316	323	330	337	344	351	358	365	372
622	379	386	393	400	407	414	421	428	435	442
623	449	456	463	470	477	484	491	498	505	511
624	518	525	532	539	546	553	560	567	574	581
625	588	595	602	609	616	623	630	637	644	650
626	657	664	671	678	685	692	699	706	713	720
627	727	734	741	748	754	761	768	775	782	789
628	796	803	810	817	824	831	837	844	851	858
629	865	872	879	886	893	900	906	913	920	927
630	934	941	948	955	962	969	975	982	989	996
631	80 003	010	017	024	030	037	044	051	058	065
632	072	079	085	092	099	106	113	120	127	134
633	140	147	154	161	168	175	182	188	195	202
634	209	216	223	229	236	243	250	257	264	271
635	277	284	291	298	305	312	318	325	332	339
636	346	353	359	366	373	380	387	393	400	407
637	414	421	428	434	441	448	455	462	468	475
638	482	489	496	502	509	516	523	530	536	543
639	550	557	564	570	577	584	591	598	604	611
640	618	625	632	638	645	652	659	665	672	679
641	686	693	699	706	713	720	726	733	740	747
642	754	760	767	774	781	787	794	801	808	814
643	821	828	835	841	848	855	862	868	875	882
644	889	895	902	909	916	922	929	936	943	949
645	956	963	969	976	983	990	996	*003	*010	*017
646	81 023	030	037	043	050	057	064	070	077	084
647	090	097	104	111	117	124	131	137	144	151
648	158	164	171	178	184	191	198	204	211	218
649	224	231	238	245	251	258	265	271	278	285
650	291	298	305	311	318	325	331	338	345	351
N.	L. 0	1	2	3	4	5	6	7	8	9

P. P.

8

1	0.8
2	1.6
3	2.4
4	3.2
5	4.0
6	4.8
7	5.6
8	6.4
9	7.2

7

1	0.7
2	1.4
3	2.1
4	2.8
5	3.5
6	4.2
7	4.9
8	5.6
9	6.3

6

1	0.6
2	1.2
3	1.8
4	2.4
5	3.0
6	3.6
7	4.2
8	4.8
9	5.4

N.	L. 0	1	2	3	4	5	6	7	8	9	P. P.
650	81 291	298	305	311	318	325	331	338	345	351	
651	358	365	371	378	385	391	398	405	411	418	
652	425	431	438	445	451	458	465	471	478	485	
653	491	498	505	511	518	525	531	538	544	551	
654	558	564	571	578	584	591	598	604	611	617	
655	624	631	637	644	651	657	664	671	677	684	
656	690	697	704	710	717	723	730	737	743	750	
657	757	763	770	776	783	790	796	803	809	816	
658	823	829	836	842	849	856	862	869	875	882	
659	889	895	902	908	915	921	928	935	941	948	
660	954	961	968	974	981	987	994	*000	*007	*014	**7**
661	82 020	027	033	040	046	053	060	066	073	079	1 0.7
662	086	092	099	105	112	119	125	132	138	145	2 1.4
663	151	158	164	171	178	184	191	197	204	210	3 2.1
664	217	223	230	236	243	249	256	263	269	276	4 2.8
665	282	289	295	302	308	315	321	328	334	341	5 3.5
666	347	354	360	367	373	380	387	393	400	406	6 4.2
667	413	419	426	432	439	445	452	458	465	471	7 4.9
668	478	484	491	497	504	510	517	523	530	536	8 5.6
669	543	549	556	562	569	575	582	588	595	601	9 6.3
670	607	614	620	627	633	640	646	653	659	666	
671	672	679	685	692	698	705	711	718	724	730	
672	737	743	750	756	763	769	776	782	789	795	
673	802	808	814	821	827	834	840	847	853	860	
674	866	872	879	885	892	898	905	911	918	924	
675	930	937	943	950	956	963	969	975	982	988	
676	995	*001	*008	*014	*020	*027	*033	*040	*046	*052	
677	83 059	065	072	078	085	091	097	104	110	117	
678	123	129	136	142	149	155	161	168	174	181	
679	187	193	200	206	213	219	225	232	238	245	
680	251	257	264	270	276	283	289	296	302	308	
681	315	321	327	334	340	347	353	359	366	372	
682	378	385	391	398	404	410	417	423	429	436	
683	442	448	455	461	467	474	480	487	493	499	
684	506	512	518	525	531	537	544	550	556	563	
685	569	575	582	588	594	601	607	613	620	626	**6.**
686	632	639	645	651	658	664	670	677	683	689	1 0.6
687	696	702	708	715	721	727	734	740	746	753	2 1.2
688	759	765	771	778	784	790	797	803	809	816	3 1.8
689	822	828	835	841	847	853	860	866	872	879	4 2.4
690	885	891	897	904	910	916	923	929	935	942	5 3.0
691	948	954	960	967	973	979	985	992	998	*004	6 3.6
692	84 011	017	023	029	036	042	048	055	061	067	7 4.2
693	073	080	086	092	098	105	111	117	123	130	8 4.8
694	136	142	148	155	161	167	173	180	186	192	9 5.4
695	198	205	211	217	223	230	236	242	248	255	
696	261	267	273	280	286	292	298	305	311	317	
697	323	330	336	342	348	354	361	367	373	379	
698	386	392	398	404	410	417	423	429	435	442	
699	448	454	460	466	473	479	485	491	497	504	
700	510	516	522	528	535	541	547	553	559	566	
N.	L. 0	1	2	3	4	5	6	7	8	9	P. P.

N.	L. 0	1	2	3	4	5	6	7	8	9
700	84 510	516	522	528	535	541	547	553	559	566
701	572	578	584	590	597	603	609	615	621	628
702	634	640	646	652	658	665	671	677	683	689
703	696	702	708	714	720	726	733	739	745	751
704	757	763	770	776	782	788	794	800	807	813
705	819	825	831	837	844	850	856	862	868	874
706	880	887	893	899	905	911	917	924	930	936
707	942	948	954	960	967	973	979	985	991	997
708	85 003	009	016	022	028	034	040	046	052	058
709	065	071	077	083	089	095	101	107	114	120
710	126	132	138	144	150	156	163	169	175	181
711	187	193	199	205	211	217	224	230	236	242
712	248	254	260	266	272	278	285	291	297	303
713	309	315	321	327	333	339	345	352	358	364
714	370	376	382	388	394	400	406	412	418	425
715	431	437	443	449	455	461	467	473	479	485
716	491	497	503	509	516	522	528	534	540	546
717	552	558	564	570	576	582	588	594	600	606
718	612	618	625	631	637	643	649	655	661	667
719	673	679	685	691	697	703	709	715	721	727
720	733	739	745	751	757	763	769	775	781	788
721	794	800	806	812	818	824	830	836	842	848
722	854	860	866	872	878	884	890	896	902	908
723	914	920	926	932	938	944	950	956	962	968
724	974	980	986	992	998	*004	*010	*016	*022	*028
725	86 034	040	046	052	058	064	070	076	082	088
726	094	100	106	112	118	124	130	136	141	147
727	153	159	165	171	177	183	189	195	201	207
728	213	219	225	231	237	243	249	255	261	267
729	273	279	285	291	297	303	308	314	320	326
730	332	338	344	350	356	362	368	374	380	386
731	392	398	404	410	415	421	427	433	439	445
732	451	457	463	469	475	481	487	493	499	504
733	510	516	522	528	534	540	546	552	558	564
734	570	576	581	587	593	599	605	611	617	623
735	629	635	641	646	652	658	664	670	676	682
736	688	694	700	705	711	717	723	729	735	741
737	747	753	759	764	770	776	782	788	794	800
738	806	812	817	823	829	835	841	847	853	859
739	864	870	876	882	888	894	900	906	911	917
740	923	929	935	941	947	953	958	964	970	976
741	982	988	994	999	*005	*011	*017	*023	*029	*035
742	87 040	046	052	058	064	070	075	081	087	093
743	099	105	111	116	122	128	134	140	146	151
744	157	163	169	175	181	186	192	198	204	210
745	216	221	227	233	239	245	251	256	262	268
746	274	280	286	291	297	303	309	315	320	326
747	332	338	344	349	355	361	367	373	379	384
748	390	396	402	408	413	419	425	431	437	442
749	448	454	460	466	471	477	483	489	495	500
750	506	512	518	523	529	535	541	547	552	558
N.	L. 0	1	2	3	4	5	6	7	8	9

P. P.

7
1	0.7
2	1.4
3	2.1
4	2.8
5	3.5
6	4.2
7	4.9
8	5.6
9	6.3

6
1	0.6
2	1.2
3	1.8
4	2.4
5	3.0
6	3.6
7	4.2
8	4.8
9	5.4

5
1	0.5
2	1.0
3	1.5
4	2.0
5	2.5
6	3.0
7	3.5
8	4.0
9	4.5

N.	L. 0	1	2	3	4	5	6	7	8	9	P. P.
750	87 506	512	518	523	529	535	541	547	552	558	
751	564	570	576	581	587	593	599	604	610	616	
752	622	628	633	639	645	651	656	662	668	674	
753	679	685	691	697	703	708	714	720	726	731	
754	737	743	749	754	760	766	772	777	783	789	
755	795	800	806	812	818	823	829	835	841	846	
756	852	858	864	869	875	881	887	892	898	904	
757	910	915	921	927	933	938	944	950	955	961	
758	967	973	978	984	990	996	*001	*007	*013	*018	
759	88 024	030	036	041	047	053	058	064	070	076	
760	081	087	093	098	104	110	116	121	127	133	
761	138	144	150	156	161	167	173	178	184	190	
762	195	201	207	213	218	224	230	235	241	247	
763	252	258	264	270	275	281	287	292	298	304	
764	309	315	321	326	332	338	343	349	355	360	
765	366	372	377	383	389	395	400	406	412	417	
766	423	429	434	440	446	451	457	463	468	474	
767	480	485	491	497	502	508	513	519	525	530	
768	536	542	547	553	559	564	570	576	581	587	
769	593	598	604	610	615	621	627	632	638	643	
770	649	655	660	666	672	677	683	689	694	700	
771	705	711	717	722	728	734	739	745	750	756	
772	762	767	773	779	784	790	795	801	807	812	
773	818	824	829	835	840	846	852	857	863	868	
774	874	880	885	891	897	902	908	913	919	925	
775	930	936	941	947	953	958	964	969	975	981	
776	986	992	997	*003	*009	*014	*020	*025	*031	*037	
777	89 042	048	053	059	064	070	076	081	087	092	
778	098	104	109	115	120	126	131	137	143	148	
779	154	159	165	170	176	182	187	193	198	204	
780	209	215	221	226	232	237	243	248	254	260	
781	265	271	276	282	287	293	298	304	310	315	
782	321	326	332	337	343	348	354	360	365	371	
783	376	382	387	393	398	404	409	415	421	426	
784	432	437	443	448	454	459	465	470	476	481	
785	487	492	498	504	509	515	520	526	531	537	
786	542	548	553	559	564	570	575	581	586	592	
787	597	603	609	614	620	625	631	636	642	647	
788	653	658	664	669	675	680	686	691	697	702	
789	708	.713	719	724	730	735	741	746	752	757	
790	763	768	774	779	785	790	796	801	807	812	
791	818	823	829	834	840	845	851	856	862	867	
792	873	878	883	889	894	900	905	911	916	922	
793	927	933	938	944	949	955	960	966	971	977	
794	982	988	993	998	*004	*009	*015	*020	*026	*031	
795	90 037	042	048	053	059	064	069	075	080	086	
796	091	097	102	108	113	119	124	129	135	140	
797	146	151	157	162	168	173	179	184	189	195	
798	200	206	211	217	222	227	233	238	244	249	
799	255	260	266	271	276	282	287	293	298	304	
800	309	314	320	325	331	336	342	347	352	358	
N.	L. 0	1	2	3	4	5	6	7	8	9	P. P.

P. P.

6
1 | 0.6
2 | 1.2
3 | 1.8
4 | 2.4
5 | 3.0
6 | 3.6
7 | 4.2
8 | 4.8
9 | 5.4

5
1 | 0.5
2 | 1.0
3 | 1.5
4 | 2.0
5 | 2.5
6 | 3.0
7 | 3.5
8 | 4.0
9 | 4.5

N.	L. 0	1	2	3	4	5	6	7	8	9
800	90 309	314	320	325	331	336	342	347	352	358
801	363	369	374	380	385	390	396	401	407	412
802	417	423	428	434	439	445	450	455	461	466
803	472	477	482	488	493	499	504	509	515	520
804	526	531	536	542	547	553	558	563	569	574
805	580	585	590	596	601	607	612	617	623	628
806	634	639	644	650	655	660	666	671	677	682
807	687	693	698	703	709	714	720	725	730	736
808	741	747	752	757	763	768	773	779	784	789
809	795	800	806	811	816	822	827	832	838	843
810	849	854	859	865	870	875	881	886	891	897
811	902	907	913	918	924	929	934	940	945	950
812	956	961	966	972	977	982	988	993	998	*004
813	91 009	014	020	025	030	036	041	046	052	057
814	062	068	073	078	084	089	094	100	105	110
815	116	121	126	132	137	142	148	153	158	164
816	169	174	180	185	190	196	201	206	212	217
817	222	228	233	238	243	249	254	259	265	270
818	275	281	286	291	297	302	307	312	318	323
819	328	334	339	344	350	355	360	365	371	376
820	381	387	392	397	403	408	413	418	424	429
821	434	440	445	450	455	461	466	471	477	482
822	487	492	498	503	508	514	519	524	529	535
823	540	545	551	556	561	566	572	577	582	587
824	593	598	603	609	614	619	624	630	635	640
825	645	651	656	661	666	672	677	682	687	693
826	698	703	709	714	719	724	730	735	740	745
827	751	756	761	766	772	777	782	787	793	798
828	803	808	814	819	824	829	834	840	845	850
829	855	861	866	871	876	882	887	892	897	903
830	908	913	918	924	929	934	939	944	950	955
831	960	965	971	976	981	986	991	997	*002	*007
832	92 012	018	023	028	033	038	044	049	054	059
833	065	070	075	080	085	091	096	101	106	111
834	117	122	127	132	137	143	148	153	158	163
835	169	174	179	184	189	195	200	205	210	215
836	221	226	231	236	241	247	252	257	262	267
837	273	278	283	288	293	298	304	309	314	319
838	324	330	335	340	345	350	355	361	366	371
839	376	381	387	392	397	402	407	412	418	423
840	428	433	438	443	449	454	459	464	469	474
841	480	485	490	495	500	505	511	516	521	526
842	531	536	542	547	552	557	562	567	572	578
843	583	588	593	598	603	609	614	619	624	629
844	634	639	645	650	655	660	665	670	675	681
845	686	691	696	701	706	711	716	722	727	732
846	737	742	747	752	758	763	768	773	778	783
847	788	793	799	804	809	814	819	824	829	834
848	840	845	850	855	860	865	870	875	881	886
849	891	896	901	906	911	916	921	927	932	937
850	942	947	952	957	962	967	973	978	983	988
N.	L. 0	1	2	3	4	5	6	7	8	9

P. P.

6

1	0.6
2	1.2
3	1.8
4	2.4
5	3.0
6	3.6
7	4.2
8	4.8
9	5.4

8

1	0.5
2	1.0
3	1.5
4	2.0
5	2.5
6	3.0
7	3.5
8	4.0
9	4.5

N.	L. 0	1	2	3	4	5	6	7	8	9	P. P.
850	92 942	947	952	957	962	967	973	978	983	988	
851	993	998	*003	*008	*013	*018	*024	*029	*034	*039	
852	93 044	049	054	059	064	069	075	080	085	090	
853	095	100	105	110	115	120	125	131	136	141	
854	146	151	156	161	166	171	176	181	186	192	
855	197	202	207	212	217	222	227	232	237	242	
856	247	252	258	263	268	273	278	283	288	293	
857	298	303	308	313	318	323	328	334	339	344	6
858	349	354	359	364	369	374	379	384	389	394	1 | 0.6
859	399	404	409	414	420	425	430	435	440	445	2 | 1.2
860	450	455	460	465	470	475	480	485	490	495	3 | 1.8
861	500	505	510	515	520	526	531	536	541	546	4 | 2.4
862	551	556	561	566	571	576	581	586	591	596	5 | 3.0
863	601	606	611	616	621	626	631	636	641	646	6 | 3.6
864	651	656	661	666	671	676	682	687	692	697	7 | 4.2
865	702	707	712	717	722	727	732	737	742	747	8 | 4.8
866	752	757	762	767	772	777	782	787	792	797	9 | 5.4
867	802	807	812	817	822	827	832	837	842	847	
868	852	857	862	867	872	877	882	887	892	897	
869	902	907	912	917	922	927	932	937	942	947	
870	952	957	962	967	972	977	982	987	992	997	
871	94 002	007	012	017	022	027	032	037	042	047	
872	052	057	062	067	072	077	082	086	091	096	5.
873	101	106	111	116	121	126	131	136	141	146	1 | 0.5
874	151	156	161	166	171	176	181	186	191	196	2 | 1.0
875	201	206	211	216	221	226	231	236	240	245	3 | 1.5
876	250	255	260	265	270	275	280	285	290	295	4 | 2.0
877	300	305	310	315	320	325	330	335	340	345	5 | 2.5
878	349	354	359	364	369	374	379	384	389	394	6 | 3.0
879	399	404	409	414	419	424	429	433	438	443	7 | 3.5
880	448	453	458	463	468	473	478	483	488	493	8 | 4.0
881	498	503	507	512	517	522	527	532	537	542	9 | 4.5
882	547	552	557	562	567	571	576	581	586	591	
883	596	601	606	611	616	621	626	630	635	640	
884	645	650	655	660	665	670	675	680	685	689	
885	694	699	704	709	714	719	724	729	734	738	
886	743	748	753	758	763	768	773	778	783	787	
887	792	797	802	807	812	817	822	827	832	836	
888	841	846	851	856	861	866	871	876	880	885	4
889	890	895	900	905	910	915	919	924	929	934	1 | 0.4
890	939	944	949	954	959	963	968	973	978	983	2 | 0.8
891	988	993	998	*002	*007	*012	*017	*022	*027	*032	3 | 1.2
892	95 036	041	046	051	056	061	066	071	075	080	4 | 1.6
893	085	090	095	100	105	109	114	119	124	129	5 | 2.0
894	134	139	143	148	153	158	163	168	173	177	6 | 2.4
895	182	187	192	197	202	207	211	216	221	226	7 | 2.8
896	231	236	240	245	250	255	260	265	270	274	8 | 3.2
897	279	284	289	294	299	303	308	313	318	323	9 | 3.6
898	328	332	337	342	347	352	357	361	366	371	
899	376	381	386	390	395	400	405	410	415	419	
900	424	429	434	439	444	448	453	458	463	468	
N.	L. 0	1	2	3	4	5	6	7	8	9	P. P.

N.	L. 0	1	2	3	4	5	6	7	8	9
900	95 424	429	434	439	444	448	453	458	463	468
901	472	477	482	487	492	497	501	506	511	516
902	521	525	530	535	540	545	550	554	559	564
903	569	574	578	583	588	593	598	602	607	612
904	617	622	626	631	636	641	646	650	655	660
905	665	670	674	679	684	689	694	698	703	708
906	713	718	722	727	732	737	742	746	751	756
907	761	766	770	775	780	785	789	794	799	804
908	809	813	818	823	828	832	837	842	847	852
909	856	861	866	871	875	880	885	890	895	899
910	904	909	914	918	923	928	933	938	942	947
911	952	957	961	966	971	976	980	985	990	995
912	999	*004	*009	*014	*019	*023	*028	*033	*038	*042
913	96 047	052	057	061	066	071	076	080	085	090
914	095	099	104	109	114	118	123	128	133	137
915	142	147	152	156	161	166	171	175	180	185
916	190	194	199	204	209	213	218	223	227	232
917	237	242	246	251	256	261	265	270	275	280
918	284	289	294	298	303	308	313	317	322	327
919	332	336	341	346	350	355	360	365	369	374
920	379	384	388	393	398	402	407	412	417	421
921	426	431	435	440	445	450	454	459	464	468
922	473	478	483	487	492	497	501	506	511	515
923	520	525	530	534	539	544	548	553	558	562
924	567	572	577	581	586	591	595	600	605	609
925	614	619	624	628	633	638	642	647	652	656
926	661	666	670	675	680	685	689	694	699	703
927	708	713	717	722	727	731	736	741	745	750
928	755	759	764	769	774	778	783	788	792	797
929	802	806	811	816	820	825	830	834	839	844
930	848	853	858	862	867	872	876	881	886	890
931	895	900	904	909	914	918	923	928	932	937
932	942	946	951	956	960	965	970	974	979	984
933	988	993	997	*002	*007	*011	*016	*021	*025	*030
934	97 035	039	044	049	053	058	063	067	072	077
935	081	086	090	095	100	104	109	114	118	123
936	128	132	137	142	146	151	155	160	165	169
937	174	179	183	188	192	197	202	206	211	216
938	220	225	230	234	239	243	248	253	257	262
939	267	271	276	280	285	290	294	299	304	308
940	313	317	322	327	331	336	340	345	350	354
941	359	364	368	373	377	382	387	391	396	400
942	405	410	414	419	424	428	433	437	442	447
943	451	456	460	465	470	474	479	483	488	493
944	497	502	506	511	516	520	525	529	534	539
945	543	548	552	557	562	566	571	575	580	585
946	589	594	598	603	607	612	617	621	626	630
947	635	640	644	649	653	658	663	667	672	676
948	681	685	690	695	699	704	708	713	717	722
949	727	731	736	740	745	749	754	759	763	768
950	772	777	782	786	791	795	800	804	809	813
N.	L. 0	1	2	3	4	5	6	7	8	9

P. P.

5

1	0.5
2	1.0
3	1.5
4	2.0
5	2.5
6	3.0
7	3.5
8	4.0
9	4.5

4

1	0.4
2	0.8
3	1.2
4	1.6
5	2.0
6	2.4
7	2.8
8	3.2
9	3.6

P. P.

N.	L. 0	1	2	3	4	5	6	7	8	9
950	97 772	777	782	786	791	795	800	804	809	813
951	818	823	827	832	836	841	845	850	855	859
952	864	868	873	877	882	886	891	896	900	905
953	909	914	918	923	928	932	937	941	946	950
954	955	959	964	968	973	978	982	987	991	996
955	98 000	005	009	014	019	023	028	032	037	041
956	046	050	055	059	064	068	073	078	082	087
957	091	096	100	105	109	114	118	123	127	132
958	137	141	146	150	155	159	164	168	173	177
959	182	186	191	195	200	204	209	214	218	223
960	227	232	236	241	245	250	254	259	263	268
961	272	277	281	286	290	295	299	304	308	313
962	318	322	327	331	336	340	345	349	354	358
963	363	367	372	376	381	385	390	394	399	403
964	408	412	417	421	426	430	435	439	444	448
965	453	457	462	466	471	475	480	484	489	493
966	498	502	507	511	516	520	525	529	534	538
967	543	547	552	556	561	565	570	574	579	583
968	588	592	597	601	605	610	614	619	623	628
969	632	637	641	646	650	655	659	664	668	673
970	677	682	686	691	695	700	704	709	713	717
971	722	726	731	735	740	744	749	753	758	762
972	767	771	776	780	784	789	793	798	802	807
973	811	816	820	825	829	834	838	843	847	851
974	856	860	865	869	874	878	883	887	892	896
975	900	905	909	914	918	923	927	932	936	941
976	945	949	954	958	963	967	972	976	981	985
977	989	994	998	*003	*007	*012	*016	*021	*025	*029
978	99 034	038	043	047	052	056	061	065	069	074
979	078	083	087	092	096	100	105	109	114	118
980	123	127	131	136	140	145	149	154	158	162
981	167	171	176	180	185	189	193	198	202	207
982	211	216	220	224	229	233	238	242	247	251
983	255	260	264	269	273	277	282	286	291	295
984	300	304	308	313	317	322	326	330	335	339
985	344	348	352	357	361	366	370	374	379	383
986	388	392	396	401	405	410	414	419	423	427
987	432	436	441	445	449	454	458	463	467	471
988	476	480	484	489	493	498	502	506	511	515
989	520	524	528	533	537	542	546	550	555	559
990	564	568	572	577	581	585	590	594	599	603
991	607	612	616	621	625	629	634	638	642	647
992	651	656	660	664	669	673	677	682	686	691
993	695	699	704	708	712	717	721	726	730	734
994	739	743	747	752	756	760	765	769	774	778
995	782	787	791	795	800	804	808	813	817	822
996	826	830	835	839	843	848	852	856	861	865
997	870	874	878	883	887	891	896	900	904	909
998	913	917	922	926	930	935	939	944	948	952
999	957	961	965	970	974	978	983	987	991	996
1000	00 000	004	009	013	017	022	026	030	035	039
N.	L. 0	1	2	3	4	5	6	7	8	9

P. P.

5
1	0.5
2	1.0
3	1.5
4	2.0
5	2.5
6	3.0
7	3.5
8	4.0
9	4.5

4
1	0.4
2	0.8
3	1.2
4	1.6
5	2.0
6	2.4
7	2.8
8	3.2
9	3.6

TABLES

OF

NATURAL SINES, COSINES, TANGENTS, AND COTANGENTS

GIVING THE VALUES OF THE FUNCTIONS FOR ALL DEGREES AND MINUTES FROM
0° TO 90°

′	0° Sine	0° Cosine	1° Sine	1° Cosine	2° Sine	2° Cosine	3° Sine	3° Cosine	4° Sine	4° Cosine	′
0	.00000	1.	.01745	.99985	.03490	.99939	.05234	.99863	.06976	.99756	60
1	.00029	1.	.01774	.99984	.03519	.99938	.05263	.99861	.07005	.99754	59
2	.00058	1.	.01803	.99984	.03548	.99937	.05292	.99860	.07034	.99752	58
3	.00087	1.	.01832	.99983	.03577	.99936	.05321	.99858	.07063	.99750	57
4	.00116	1.	.01862	.99983	.03606	.99935	.05350	.99857	.07092	.99748	56
5	.00145	1.	.01891	.99982	.03635	.99934	.05379	.99855	.07121	.99746	55
6	.00175	1.	.01920	.99982	.03664	.99933	.05408	.99854	.07150	.99744	54
7	.00204	1.	.01949	.99981	.03693	.99932	.05437	.99852	.07179	.99742	53
8	.00233	1.	.01978	.99980	.03723	.99931	.05466	.99851	.07208	.99740	52
9	.00262	1.	.02007	.99980	.03752	.99930	.05495	.99849	.07237	.99738	51
10	.00291	1.	.02036	.99979	.03781	.99929	.05524	.99847	.07266	.99736	50
11	.00320	.99999	.02065	.99979	.03810	.99927	.05553	.99846	.07295	.99734	49
12	.00349	.99999	.02094	.99978	.03839	.99926	.05582	.99844	.07324	.99731	48
13	.00378	.99999	.02123	.99977	.03868	.99925	.05611	.99842	.07353	.99729	47
14	.00407	.99999	.02152	.99977	.03897	.99924	.05640	.99841	.07382	.99727	46
15	.00436	.99999	.02181	.99976	.03926	.99923	.05669	.99839	.07411	.99725	45
16	.00465	.99999	.02211	.99976	.03955	.99922	.05698	.99838	.07440	.99723	44
17	.00495	.99999	.02240	.99975	.03984	.99921	.05727	.99836	.07469	.99721	43
18	.00524	.99999	.02269	.99974	.04013	.99919	.05756	.99834	.07498	.99719	42
19	.00553	.99998	.02298	.99974	.04042	.99918	.05785	.99833	.07527	.99716	41
20	.00582	.99998	.02327	.99973	.04071	.99917	.05814	.99831	.07556	.99714	40
21	.00611	.99998	.02356	.99972	.04100	.99916	.05844	.99829	.07585	.99712	39
22	.00640	.99998	.02385	.99972	.04129	.99915	.05873	.99827	.07614	.99710	38
23	.00669	.99998	.02414	.99971	.04159	.99913	.05902	.99826	.07643	.99708	37
24	.00698	.99998	.02443	.99970	.04188	.99912	.05931	.99824	.07672	.99705	36
25	.00727	.99997	.02472	.99969	.04217	.99911	.05960	.99822	.07701	.99703	35
26	.00756	.99997	.02501	.99969	.04246	.99910	.05989	.99821	.07730	.99701	34
27	.00785	.99997	.02530	.99968	.04275	.99909	.06018	.99819	.07759	.99699	33
28	.00814	.99997	.02560	.99967	.04304	.99907	.06047	.99817	.07788	.99696	32
29	.00844	.99996	.02589	.99966	.04333	.99906	.06076	.99815	.07817	.99694	31
30	.00873	.99996	.02618	.99966	.04362	.99905	.06105	.99813	.07846	.99692	30
31	.00902	.99996	.02647	.99965	.04391	.99904	.06134	.99812	.07875	.99689	29
32	.00931	.99996	.02676	.99964	.04420	.99902	.06163	.99810	.07904	.99687	28
33	.00960	.99995	.02705	.99963	.04449	.99901	.06192	.99808	.07933	.99685	27
34	.00989	.99995	.02734	.99963	.04478	.99900	.06221	.99806	.07962	.99683	26
35	.01018	.99995	.02763	.99962	.04507	.99898	.06250	.99804	.07991	.99680	25
36	.01047	.99995	.02792	.99961	.04536	.99897	.06279	.99803	.08020	.99678	24
37	.01076	.99994	.02821	.99960	.04565	.99896	.06308	.99801	.08049	.99676	23
38	.01105	.99994	.02850	.99959	.04594	.99894	.06337	.99799	.08078	.99673	22
39	.01134	.99994	.02879	.99959	.04623	.99893	.06366	.99797	.08107	.99671	21
40	.01164	.99993	.02908	.99958	.04653	.99892	.06395	.99795	.08136	.99668	20
41	.01193	.99993	.02938	.99957	.04682	.99890	.06424	.99793	.08165	.99666	19
42	.01222	.99993	.02967	.99956	.04711	.99889	.06453	.99792	.08194	.99664	18
43	.01251	.99992	.02996	.99955	.04740	.99888	.06482	.99790	.08223	.99661	17
44	.01280	.99992	.03025	.99954	.04769	.99886	.06511	.99788	.08252	.99659	16
45	.01309	.99991	.03054	.99953	.04798	.99885	.06540	.99786	.08281	.99657	15
46	.01338	.99991	.03083	.99952	.04827	.99883	.06569	.99784	.08310	.99654	14
47	.01367	.99991	.03112	.99952	.04856	.99882	.06598	.99782	.08339	.99652	13
48	.01396	.99990	.03141	.99951	.04885	.99881	.06627	.99780	.08368	.99649	12
49	.01425	.99990	.03170	.99950	.04914	.99879	.06656	.99778	.08397	.99647	11
50	.01454	.99989	.03199	.99949	.04943	.99878	.06685	.99776	.08426	.99644	10
51	.01483	.99989	.03228	.99948	.04972	.99876	.06714	.99774	.08455	.99642	9
52	.01513	.99989	.03257	.99947	.05001	.99875	.06743	.99772	.08484	.99639	8
53	.01542	.99988	.03286	.99946	.05030	.99873	.06773	.99770	.08513	.99637	7
54	.01571	.99988	.03316	.99945	.05059	.99872	.06802	.99768	.08542	.99635	6
55	.01600	.99987	.03345	.99944	.05088	.99870	.06831	.99766	.08571	.99632	5
56	.01629	.99987	.03374	.99943	.05117	.99869	.06860	.99764	.08600	.99630	4
57	.01658	.99986	.03403	.99942	.05146	.99867	.06889	.99762	.08629	.99627	3
58	.01687	.99986	.03432	.99941	.05175	.99866	.06918	.99760	.08658	.99625	2
59	.01716	.99985	.03461	.99940	.05205	.99864	.06947	.99758	.08687	.99622	1
60	.01745	.99985	.03490	.99939	.05234	.99863	.06976	.99756	.08716	.99619	0
′	Cosine	Sine	Cosine	Sine	Cosine	Sine	Cosine	Sine	Cosine	Sine	′
	89°		88°		87°		86°		85°		

′	5° Sine	5° Cosine	6° Sine	6° Cosine	7° Sine	7° Cosine	8° Sine	8° Cosine	9° Sine	9° Cosine	′
0	.08716	.99619	.10453	.99452	.12187	.99255	.13917	.99027	.15643	.98769	60
1	.08745	.99617	.10482	.99449	.12216	.99251	.13946	.99023	.15672	.98764	59
2	.08774	.99614	.10511	.99446	.12245	.99248	.13975	.99019	.15701	.98760	58
3	.08803	.99612	.10540	.99443	.12274	.99244	.14004	.99015	.15730	.98755	57
4	.08831	.99609	.10569	.99440	.12302	.99240	.14033	.99011	.15758	.98751	56
5	.08860	.99607	.10597	.99437	.12331	.99237	.14061	.99006	.15787	.98746	55
6	.08889	.99604	.10626	.99434	.12360	.99233	.14090	.99002	.15816	.98741	54
7	.08918	.99602	.10655	.99431	.12389	.99230	.14119	.98998	.15845	.98737	53
8	.08947	.99599	.10684	.99428	.12418	.99226	.14148	.98994	.15873	.98732	52
9	.08976	.99596	.10713	.99424	.12447	.99222	.14177	.98990	.15902	.98728	51
10	.09005	.99594	.10742	.99421	.12476	.99219	.14205	.98986	.15931	.98723	50
11	.09034	.99591	.10771	.99418	.12504	.99215	.14234	.98982	.15959	.98718	49
12	.09063	.99588	.10800	.99415	.12533	.99211	.14263	.98978	.15988	.98714	48
13	.09092	.99586	.10829	.99412	.12562	.99208	.14292	.98973	.16017	.98709	47
14	.09121	.99583	.10858	.99409	.12591	.99204	.14320	.98969	.16046	.98704	46
15	.09150	.99580	.10887	.99406	.12620	.99200	.14349	.98965	.16074	.98700	45
16	.09179	.99578	.10916	.99402	.12649	.99197	.14378	.98961	.16103	.98695	44
17	.09208	.99575	.10945	.99399	.12678	.99193	.14407	.98957	.16132	.98690	43
18	.09237	.99572	.10973	.99396	.12706	.99189	.14436	.98953	.16160	.98686	42
19	.09266	.99570	.11002	.99393	.12735	.99186	.14464	.98948	.16189	.98681	41
20	.09295	.99567	.11031	.99390	.12764	.99182	.14493	.98944	.16218	.98676	40
21	.09324	.99564	.11060	.99386	.12793	.99178	.14522	.98940	.16246	.98671	39
22	.09353	.99562	.11089	.99383	.12822	.99175	.14551	.98936	.16275	.98667	38
23	.09382	.99559	.11118	.99380	.12851	.99171	.14580	.98931	.16304	.98662	37
24	.09411	.99556	.11147	.99377	.12880	.99167	.14608	.98927	.16333	.98657	36
25	.09440	.99553	.11176	.99374	.12908	.99163	.14637	.98923	.16361	.98652	35
26	.09469	.99551	.11205	.99370	.12937	.99160	.14666	.98919	.16390	.98648	34
27	.09498	.99548	.11234	.99367	.12966	.99156	.14695	.98914	.16419	.98643	33
28	.09527	.99545	.11263	.99364	.12995	.99152	.14723	.98910	.16447	.98638	32
29	.09556	.99542	.11291	.99360	.13024	.99148	.14752	.98906	.16476	.98633	31
30	.09585	.99540	.11320	.99357	.13053	.99144	.14781	.98902	.16505	.98629	30
31	.09614	.99537	.11349	.99354	.13081	.99141	.14810	.98897	.16533	.98624	29
32	.09642	.99534	.11378	.99351	.13110	.99137	.14838	.98893	.16562	.98619	28
33	.09671	.99531	.11407	.99347	.13139	.99133	.14867	.98889	.16591	.98614	27
34	.09700	.99528	.11436	.99344	.13168	.99129	.14896	.98884	.16620	.98609	26
35	.09729	.99526	.11465	.99341	.13197	.99125	.14925	.98880	.16648	.98604	25
36	.09758	.99523	.11494	.99337	.13226	.99122	.14954	.98876	.16677	.98600	24
37	.09787	.99520	.11523	.99334	.13254	.99118	.14982	.98871	.16706	.98595	23
38	.09816	.99517	.11552	.99331	.13283	.99114	.15011	.98867	.16734	.98590	22
39	.09845	.99514	.11580	.99327	.13312	.99110	.15040	.98863	.16763	.98585	21
40	.09874	.99511	.11609	.99324	.13341	.99106	.15069	.98858	.16792	.98580	20
41	.09903	.99508	.11638	.99320	.13370	.99102	.15097	.98854	.16820	.98575	19
42	.09932	.99506	.11667	.99317	.13399	.99098	.15126	.98849	.16849	.98570	18
43	.09961	.99503	.11696	.99314	.13427	.99094	.15155	.98845	.16878	.98565	17
44	.09990	.99500	.11725	.99310	.13456	.99091	.15184	.98841	.16906	.98561	16
45	.10019	.99497	.11754	.99307	.13485	.99087	.15212	.98836	.16935	.98556	15
46	.10048	.99494	.11783	.99303	.13514	.99083	.15241	.98832	.16964	.98551	14
47	.10077	.99491	.11812	.99300	.13543	.99079	.15270	.98827	.16992	.98546	13
48	.10106	.99488	.11840	.99297	.13572	.99075	.15299	.98823	.17021	.98541	12
49	.10135	.99485	.11869	.99293	.13600	.99071	.15327	.98818	.17050	.98536	11
50	.10164	.99482	.11898	.99290	.13629	.99067	.15356	.98814	.17078	.98531	10
51	.10192	.99479	.11927	.99286	.13658	.99063	.15385	.98809	.17107	.98526	9
52	.10221	.99476	.11956	.99283	.13687	.99059	.15414	.98805	.17136	.98521	8
53	.10250	.99473	.11985	.99279	.13716	.99055	.15442	.98800	.17164	.98516	7
54	.10279	.99470	.12014	.99276	.13744	.99051	.15471	.98796	.17193	.98511	6
55	.10308	.99467	.12043	.99272	.13773	.99047	.15500	.98791	.17222	.98506	5
56	.10337	.99464	.12071	.99269	.13802	.99043	.15529	.98787	.17250	.98501	4
57	.10366	.99461	.12100	.99265	.13831	.99039	.15557	.98782	.17279	.98496	3
58	.10395	.99458	.12129	.99262	.13860	.99035	.15586	.98778	.17308	.98491	2
59	.10424	.99455	.12158	.99258	.13889	.99031	.15615	.98773	.17336	.98486	1
60	.10453	.99452	.12187	.99255	.13917	.99027	.15643	.98769	.17365	.98481	0
′	Cosine	Sine	Cosine	Sine	Cosine	Sine	Cosine	Sine	Cosine	Sine	′

| | 84° | 83° | 82° | 81° | 80° | |

′	10° Sine	10° Cosine	11° Sine	11° Cosine	12° Sine	12° Cosine	13° Sine	13° Cosine	14° Sine	14° Cosine	′
0	.17365	.98481	.19081	.98163	.20791	.97815	.22495	.97437	.24192	.97030	60
1	.17393	.98476	.19109	.98157	.20820	.97809	.22523	.97430	.24220	.97023	59
2	.17422	.98471	.19138	.98152	.20848	.97803	.22552	.97424	.24249	.97015	58
3	.17451	.98466	.19167	.98146	.20877	.97797	.22580	.97417	.24277	.97008	57
4	.17479	.98461	.19195	.98140	.20905	.97791	.22608	.97411	.24305	.97001	56
5	.17508	.98455	.19224	.98135	.20933	.97784	.22637	.97404	.24333	.96994	55
6	.17537	.98450	.19252	.98129	.20962	.97778	.22665	.97398	.24362	.96987	54
7	.17565	.98445	.19281	.98124	.20990	.97772	.22693	.97391	.24390	.96980	53
8	.17594	.98440	.19309	.98118	.21019	.97766	.22722	.97384	.24418	.96973	52
9	.17623	.98435	.19338	.98112	.21047	.97760	.22750	.97378	.24446	.96966	51
10	.17651	.98430	.19366	.98107	.21076	.97754	.22778	.97371	.24474	.96959	50
11	.17680	.98425	.19395	.98101	.21104	.97748	.22807	.97365	.24503	.96952	49
12	.17708	.98420	.19423	.98096	.21132	.97742	.22835	.97358	.24531	.96945	48
13	.17737	.98414	.19452	.98090	.21161	.97735	.22863	.97351	.24559	.96937	47
14	.17766	.98409	.19481	.98084	.21189	.97729	.22892	.97345	.24587	.96930	46
15	.17794	.98404	.19509	.98079	.21218	.97723	.22920	.97338	.24615	.96923	45
16	.17823	.98399	.19538	.98073	.21246	.97717	.22948	.97331	.24644	.96916	44
17	.17852	.98394	.19566	.98067	.21275	.97711	.22977	.97325	.24672	.96909	43
18	.17880	.98389	.19595	.98061	.21303	.97705	.23005	.97318	.24700	.96902	42
19	.17909	.98383	.19623	.98056	.21331	.97698	.23033	.97311	.24728	.96894	41
20	.17937	.98378	.19652	.98050	.21360	.97692	.23062	.97304	.24756	.96887	40
21	.17966	.98373	.19680	.98044	.21388	.97686	.23090	.97298	.24784	.96880	39
22	.17995	.98368	.19709	.98039	.21417	.97680	.23118	.97291	.24813	.96873	38
23	.18023	.98362	.19737	.98031	.21445	.97673	.23146	.97284	.24841	.96866	37
24	.18052	.98357	.19766	.98027	.21474	.97667	.23175	.97278	.24869	.96858	36
25	.18081	.98352	.19794	.98021	.21502	.97661	.23203	.97271	.24897	.96851	35
26	.18109	.98347	.19823	.98016	.21530	.97655	.23231	.97264	.24925	.96844	34
27	.18138	.98341	.19851	.98010	.21559	.97648	.23260	.97257	.24954	.96837	33
28	.18166	.98336	.19880	.98004	.21587	.97642	.23288	.97251	.24982	.96829	32
29	.18195	.98331	.19908	.97998	.21616	.97636	.23316	.97244	.25010	.96822	31
30	.18224	.98325	.19937	.97992	.21644	.97630	.23345	.97237	.25038	.96815	30
31	.18252	.98320	.19965	.97987	.21672	.97623	.23373	.97230	.25066	.96807	29
32	.18281	.98315	.19994	.97981	.21701	.97617	.23401	.97223	.25094	.96800	28
33	.18309	.98310	.20022	.97975	.21729	.97611	.23429	.97217	.25122	.96793	27
34	.18338	.98304	.20051	.97969	.21758	.97604	.23458	.97210	.25151	.96786	26
35	.18367	.98299	.20079	.97963	.21786	.97598	.23486	.97203	.25179	.96778	25
36	.18395	.98294	.20108	.97958	.21814	.97592	.23514	.97196	.25207	.96771	24
37	.18424	.98288	.20136	.97952	.21843	.97585	.23542	.97189	.25235	.96764	23
38	.18452	.98283	.20165	.97946	.21871	.97579	.23571	.97182	.25263	.96756	22
39	.18481	.98277	.20193	.97940	.21899	.97573	.23599	.97176	.25291	.96749	21
40	.18509	.98272	.20222	.97934	.21928	.97566	.23627	.97169	.25320	.96742	20
41	.18538	.98267	.20250	.97928	.21956	.97560	.23656	.97162	.25348	.96734	19
42	.18567	.98261	.20279	.97922	.21985	.97553	.23684	.97155	.25376	.96727	18
43	.18595	.98256	.20307	.97916	.22013	.97547	.23712	.97148	.25404	.96719	17
44	.18624	.98250	.20336	.97910	.22041	.97541	.23740	.97141	.25432	.96712	16
45	.18652	.98245	.20364	.97905	.22070	.97534	.23769	.97134	.25460	.96705	15
46	.18681	.98240	.20393	.97899	.22098	.97528	.23797	.97127	.25488	.96697	14
47	.18710	.98234	.20421	.97893	.22126	.97521	.23825	.97120	.25516	.96690	13
48	.18738	.98229	.20450	.97887	.22155	.97515	.23853	.97113	.25545	.96682	12
49	.18767	.98223	.20478	.97881	.22183	.97508	.23882	.97106	.25573	.96675	11
50	.18795	.98218	.20507	.97875	.22212	.97502	.23910	.97100	.25601	.96667	10
51	.18824	.98212	.20535	.97869	.22240	.97496	.23938	.97093	.25629	.96660	9
52	.18852	.98207	.20563	.97863	.22268	.97489	.23966	.97086	.25657	.96653	8
53	.18881	.98201	.20592	.97857	.22297	.97483	.23995	.97079	.25685	.96645	7
54	.18910	.98196	.20620	.97851	.22325	.97476	.24023	.97072	.25713	.96638	6
55	.18938	.98190	.20649	.97845	.22353	.97470	.24051	.97065	.25741	.96630	5
56	.18967	.98185	.20677	.97839	.22382	.97463	.24079	.97058	.25769	.96623	4
57	.18995	.98179	.20706	.97833	.22410	.97457	.24108	.97051	.25798	.96615	3
58	.19024	.98174	.20734	.97827	.22438	.97450	.24136	.97044	.25826	.96608	2
59	.19052	.98168	.20763	.97821	.22467	.97444	.24164	.97037	.25854	.96600	1
60	.19081	.98163	.20791	.97815	.22495	.97437	.24192	.97030	.25882	.96593	0
′	Cosine	Sine	Cosine	Sine	Cosine	Sine	Cosine	Sine	Cosine	Sine	′
	79°		78°		77°		76°		75°		

′	15°		16°		17°		18°		19°		′
	Sine	Cosine	Sine	Cosine	Sine	Cosine	Sine	Cosine	Sine	Cosine	
0	.25882	.96593	.27564	.96126	.29237	.95630	.30902	.95106	.32557	.94552	60
1	.25910	.96585	.27592	.96118	.29265	.95622	.30929	.95097	.32584	.94542	59
2	.25938	.96578	.27620	.96110	.29293	.95613	.30957	.95088	.32612	.94533	58
3	.25966	.96570	.27648	.96102	.29321	.95605	.30985	.95079	.32639	.94523	57
4	.25994	.96562	.27676	.96094	.29348	.95596	.31012	.95070	.32667	.94514	56
5	.26022	.96555	.27704	.96086	.29376	.95588	.31040	.95061	.32694	.94504	55
6	.26050	.96547	.27731	.96078	.29404	.95579	.31068	.95052	.32722	.94495	54
7	.26079	.96540	.27759	.96070	.29432	.95571	.31095	.95043	.32749	.94485	53
8	.26107	.96532	.27787	.96062	.29460	.95562	.31123	.95033	.32777	.94476	52
9	.26135	.96524	.27815	.96054	.29487	.95554	.31151	.95024	.32804	.94466	51
10	.26163	.96517	.27843	.96046	.29515	.95545	.31178	.95015	.32832	.94457	50
11	.26191	.96509	.27871	.96037	.29543	.95536	.31206	.95006	.32859	.94447	49
12	.26219	.96502	.27899	.96029	.29571	.95528	.31233	.94997	.32887	.94438	48
13	.26247	.96494	.27927	.96021	.29599	.95519	.31261	.94988	.32914	.94428	47
14	.26275	.96486	.27955	.96013	.29626	.95511	.31289	.94979	.32942	.94418	46
15	.26303	.96479	.27983	.96005	.29654	.95502	.31316	.94970	.32969	.94409	45
16	.26331	.96471	.28011	.95997	.29682	.95493	.31344	.94961	.32997	.94399	44
17	.26359	.96463	.28039	.95989	.29710	.95485	.31372	.94952	.33024	.94390	43
18	.26387	.96456	.28067	.95981	.29737	.95476	.31399	.94943	.33051	.94380	42
19	.26415	.96448	.28095	.95972	.29765	.95467	.31427	.94933	.33079	.94370	41
20	.26443	.96440	.28123	.95964	.29793	.95459	.31454	.94924	.33106	.94361	40
21	.26471	.96433	.28150	.95956	.29821	.95450	.31482	.94915	.33134	.94351	39
22	.26500	.96425	.28178	.95948	.29849	.95441	.31510	.94906	.33161	.94342	38
23	.26528	.96417	.28206	.95940	.29876	.95433	.31537	.94897	.33189	.94332	37
24	.26556	.96410	.28234	.95931	.29904	.95424	.31565	.94888	.33216	.94322	36
25	.26584	.96402	.28262	.95923	.29932	.95415	.31593	.94878	.33244	.94313	35
26	.26612	.96394	.28290	.95915	.29960	.95407	.31620	.94869	.33271	.94303	34
27	.26640	.96386	.28318	.95907	.29987	.95398	.31648	.94860	.33298	.94293	33
28	.26668	.96379	.28346	.95898	.30015	.95389	.31675	.94851	.33326	.94284	32
29	.26696	.96371	.28374	.95890	.30043	.95380	.31703	.94842	.33353	.94274	31
30	.26724	.96363	.28402	.95882	.30071	.95372	.31730	.94832	.33381	.94264	30
31	.26752	.96355	.28429	.95874	.30098	.95363	.31758	.94823	.33408	.94254	29
32	.26780	.96347	.28457	.95865	.30126	.95354	.31786	.94814	.33436	.94245	28
33	.26808	.96340	.28485	.95857	.30154	.95345	.31813	.94805	.33463	.94235	27
34	.26836	.96332	.28513	.95849	.30182	.95337	.31841	.94795	.33490	.94225	26
35	.26864	.96324	.28541	.95841	.30209	.95328	.31868	.94786	.33518	.94215	25
36	.26892	.96316	.28569	.95832	.30237	.95319	.31896	.94777	.33545	.94206	24
37	.26920	.96308	.28597	.95824	.30265	.95310	.31923	.94768	.33573	.94196	23
38	.26948	.96301	.28625	.95816	.30292	.95301	.31951	.94758	.33600	.94186	22
39	.26976	.96293	.28652	.95807	.30320	.95293	.31979	.94749	.33627	.94176	21
40	.27004	.96285	.28680	.95799	.30348	.95284	.32006	.94740	.33655	.94167	20
41	.27032	.96277	.28708	.95791	.30376	.95275	.32034	.94730	.33682	.94157	19
42	.27060	.96269	.28736	.95782	.30403	.95266	.32061	.94721	.33710	.94147	18
43	.27088	.96261	.28764	.95774	.30431	.95257	.32089	.94712	.33737	.94137	17
44	.27116	.96253	.28792	.95766	.30459	.95248	.32116	.94702	.33764	.94127	16
45	.27144	.96246	.28820	.95757	.30486	.95240	.32144	.94693	.33792	.94118	15
46	.27172	.96238	.28847	.95749	.30514	.95231	.32171	.94684	.33819	.94108	14
47	.27200	.96230	.28875	.95740	.30542	.95222	.32199	.94674	.33846	.94098	13
48	.27228	.96222	.28903	.95732	.30570	.95213	.32227	.94665	.33874	.94088	12
49	.27256	.96214	.28931	.95724	.30597	.95204	.32254	.94656	.33901	.94078	11
50	.27284	.96206	.28959	.95715	.30625	.95195	.32282	.94646	.33929	.94068	10
51	.27312	.96198	.28987	.95707	.30653	.95186	.32309	.94637	.33956	.94058	9
52	.27340	.96190	.29015	.95698	.30680	.95177	.32337	.94627	.33983	.94049	8
53	.27368	.96182	.29042	.95690	.30708	.95168	.32364	.94618	.34011	.94039	7
54	.27396	.96174	.29070	.95681	.30736	.95159	.32392	.94609	.34038	.94029	6
55	.27424	.96166	.29098	.95673	.30763	.95150	.32419	.94599	.34065	.94019	5
56	.27452	.96158	.29126	.95664	.30791	.95142	.32447	.94590	.34093	.94009	4
57	.27480	.96150	.29154	.95656	.30819	.95133	.32474	.94580	.34120	.93999	3
58	.27508	.96142	.29182	.95647	.30846	.95124	.32502	.94571	.34147	.93989	2
59	.27536	.96134	.29209	.95639	.30874	.95115	.32529	.94561	.34175	.93979	1
60	.27564	.96126	.29237	.95630	.30902	.95106	.32557	.94552	.34202	.93969	0
′	Cosine	Sine	Cosine	Sine	Cosine	Sine	Cosine	Sine	Cosine	Sine	′
	74°		73°		72°		71°		70°		

′	20° Sine	20° Cosine	21° Sine	21° Cosine	22° Sine	22° Cosine	23° Sine	23° Cosine	24° Sine	24° Cosine	′
0	.34202	.93969	.35837	.93358	.37461	.92718	.39073	.92050	.40674	.91355	60
1	.34229	.93959	.35864	.93348	.37488	.92707	.39100	.92039	.40700	.91343	59
2	.34257	.93949	.35891	.93337	.37515	.92697	.39127	.92028	.40727	.91331	58
3	.34284	.93939	.35918	.93327	.37542	.92686	.39153	.92016	.40753	.91319	57
4	.34311	.93929	.35945	.93316	.37569	.92675	.39180	.92005	.40780	.91307	56
5	.34339	.93919	.35973	.93306	.37595	.92664	.39207	.91994	.40806	.91295	55
6	.34366	.93909	.36000	.93295	.37622	.92653	.39234	.91982	.40833	.91283	54
7	.34393	.93899	.36027	.93285	.37649	.92642	.39260	.91971	.40860	.91272	53
8	.34421	.93889	.36054	.93274	.37676	.92631	.39287	.91959	.40886	.91260	52
9	.34448	.93879	.36081	.93264	.37703	.92620	.39314	.91948	.40913	.91248	51
10	.34475	.93869	.36108	.93253	.37730	.92609	.39341	.91936	.40939	.91236	50
11	.34503	.93859	.36135	.93243	.37757	.92598	.39367	.91925	.40966	.91224	49
12	.34530	.93849	.36162	.93232	.37784	.92587	.39394	.91914	.40992	.91212	48
13	.34557	.93839	.36190	.93222	.37811	.92576	.39421	.91902	.41019	.91200	47
14	.34584	.93829	.36217	.93211	.37838	.92565	.39448	.91891	.41045	.91188	46
15	.34612	.93819	.36244	.93201	.37865	.92554	.39474	.91879	.41072	.91176	45
16	.34639	.93809	.36271	.93190	.37892	.92543	.39501	.91868	.41098	.91164	44
17	.34666	.93799	.36298	.93180	.37919	.92532	.39528	.91856	.41125	.91152	43
18	.34694	.93789	.36325	.93169	.37946	.92521	.39555	.91845	.41151	.91140	42
19	.34721	.93779	.36352	.93159	.37973	.92510	.39581	.91833	.41178	.91128	41
20	.34748	.93769	.36379	.93148	.37999	.92499	.39608	.91822	.41204	.91116	40
21	.34775	.93759	.36406	.93137	.38026	.92488	.39635	.91810	.41231	.91104	39
22	.34803	.93748	.36434	.93127	.38053	.92477	.39661	.91799	.41257	.91092	38
23	.34830	.93738	.36461	.93116	.38080	.92466	.39688	.91787	.41284	.91080	37
24	.34857	.93728	.36488	.93106	.38107	.92455	.39715	.91775	.41310	.91068	36
25	.34884	.93718	.36515	.93095	.38134	.92444	.39741	.91764	.41337	.91056	35
26	.34912	.93708	.36542	.93084	.38161	.92432	.39768	.91752	.41363	.91044	34
27	.34939	.93698	.36569	.93074	.38188	.92421	.39795	.91741	.41390	.91032	33
28	.34966	.93688	.36596	.93063	.38215	.92410	.39822	.91729	.41416	.91020	32
29	.34993	.93677	.36623	.93052	.38241	.92399	.39848	.91718	.41443	.91008	31
30	.35021	.93667	.36650	.93042	.38268	.92388	.39875	.91706	.41469	.90996	30
31	.35048	.93657	.36677	.93031	.38295	.92377	.39902	.91694	.41496	.90984	29
32	.35075	.93647	.36704	.93020	.38322	.92366	.39928	.91683	.41522	.90972	28
33	.35102	.93637	.36731	.93010	.38349	.92355	.39955	.91671	.41549	.90960	27
34	.35130	.93626	.36758	.92999	.38376	.92343	.39982	.91660	.41575	.90948	26
35	.35157	.93616	.36785	.92988	.38403	.92332	.40008	.91648	.41602	.90936	25
36	.35184	.93606	.36812	.92978	.38430	.92321	.40035	.91636	.41628	.90924	24
37	.35211	.93596	.36839	.92967	.38456	.92310	.40062	.91625	.41655	.90911	23
38	.35239	.93585	.36867	.92956	.38483	.92299	.40088	.91613	.41681	.90899	22
39	.35266	.93575	.36894	.92945	.38510	.92287	.40115	.91601	.41707	.90887	21
40	.35293	.93565	.36921	.92935	.38537	.92276	.40141	.91590	.41734	.90875	20
41	.35320	.93555	.36948	.92924	.38564	.92265	.40168	.91578	.41760	.90863	19
42	.35347	.93544	.36975	.92913	.38591	.92254	.40195	.91566	.41787	.90851	18
43	.35375	.93534	.37002	.92902	.38617	.92243	.40221	.91555	.41813	.90839	17
44	.35402	.93524	.37029	.92892	.38644	.92231	.40248	.91543	.41840	.90826	16
45	.35429	.93514	.37056	.92881	.38671	.92220	.40275	.91531	.41866	.90814	15
46	.35456	.93503	.37083	.92870	.38698	.92209	.40301	.91519	.41892	.90802	14
47	.35484	.93493	.37110	.92859	.38725	.92198	.40328	.91508	.41919	.90790	13
48	.35511	.93483	.37137	.92849	.38752	.92186	.40355	.91496	.41945	.90778	12
49	.35538	.93472	.37164	.92838	.38778	.92175	.40381	.91484	.41972	.90766	11
50	.35565	.93462	.37191	.92827	.38805	.92164	.40408	.91472	.41998	.90753	10
51	.35592	.93452	.37218	.92816	.38832	.92152	.40434	.91461	.42024	.90741	9
52	.35619	.93441	.37245	.92805	.38859	.92141	.40461	.91449	.42051	.90729	8
53	.35647	.93431	.37272	.92794	.38886	.92130	.40488	.91437	.42077	.90717	7
54	.35674	.93420	.37299	.92784	.38912	.92119	.40514	.91425	.42104	.90704	6
55	.35701	.93410	.37326	.92773	.38939	.92107	.40541	.91414	.42130	.90692	5
56	.35728	.93400	.37353	.92762	.38966	.92096	.40567	.91402	.42156	.90680	4
57	.35755	.93389	.37380	.92751	.38993	.92085	.40594	.91390	.42183	.90668	3
58	.35782	.93379	.37407	.92740	.39020	.92073	.40621	.91378	.42209	.90655	2
59	.35810	.93368	.37434	.92729	.39046	.92062	.40647	.91366	.42235	.90643	1
60	.35837	.93358	.37461	.92718	.39073	.92050	.40674	.91355	.42262	.90631	0
′	Cosine	Sine	Cosine	Sine	Cosine	Sine	Cosine	Sine	Cosine	Sine	′
	69°		68°		67°		66°		65°		

′	25°		26°		27°		28°		29°		′
	Sine	Cosine	Sine	Cosine	Sine	Cosine	Sine	Cosine	Sine	Cosine	
0	.42262	.90631	.43837	.89879	.45399	.89101	.46947	.88295	.48481	.87462	60
1	.42288	.90618	.43863	.89867	.45425	.89087	.46973	.88281	.48506	.87448	59
2	.42315	.90606	.43889	.89854	.45451	.89074	.46999	.88267	.48532	.87434	58
3	.42341	.90594	.43916	.89841	.45477	.89061	.47024	.88254	.48557	.87420	57
4	.42367	.90582	.43942	.89828	.45503	.89048	.47050	.88240	.48583	.87406	56
5	.42394	.90569	.43968	.89816	.45529	.89035	.47076	.88226	.48608	.87391	55
6	.42420	.90557	.43994	.89803	.45554	.89021	.47101	.88213	.48634	.87377	54
7	.42446	.90545	.44020	.89790	.45580	.89008	.47127	.88199	.48659	.87363	53
8	.42473	.90532	.44046	.89777	.45606	.88995	.47153	.88185	.48684	.87349	52
9	.42499	.90520	.44072	.89764	.45632	.88981	.47178	.88172	.48710	.87335	51
10	.42525	.90507	.44098	.89752	.45658	.88968	.47204	.88158	.48735	.87321	50
11	.42552	.90495	.44124	.89739	.45684	.88955	.47229	.88144	.48761	.87306	49
12	.42578	.90483	.44151	.89726	.45710	.88942	.47255	.88130	.48786	.87292	48
13	.42604	.90470	.44177	.89713	.45736	.88928	.47281	.88117	.48811	.87278	47
14	.42631	.90458	.44203	.89700	.45762	.88915	.47306	.88103	.48837	.87264	46
15	.42657	.90446	.44229	.89687	.45787	.88902	.47332	.88089	.48862	.87250	45
16	.42683	.90433	.44255	.89674	.45813	.88888	.47358	.88075	.48888	.87235	44
17	.42709	.90421	.44281	.89662	.45839	.88875	.47383	.88062	.48913	.87221	43
18	.42736	.90408	.44307	.89649	.45865	.88862	.47409	.88048	.48938	.87207	42
19	.42762	.90396	.44333	.89636	.45891	.88848	.47434	.88034	.48964	.87193	41
20	.42788	.90383	.44359	.89623	.45917	.88835	.47460	.88020	.48989	.87178	40
21	.42815	.90371	.44385	.89610	.45942	.88822	.47486	.88006	.49014	.87164	39
22	.42841	.90358	.44411	.89597	.45968	.88808	.47511	.87993	.49040	.87150	38
23	.42867	.90346	.44437	.89584	.45994	.88795	.47537	.87979	.49065	.87136	37
24	.42894	.90334	.44464	.89571	.46020	.88782	.47562	.87965	.49090	.87121	36
25	.42920	.90321	.44490	.89558	.46046	.88768	.47588	.87951	.49116	.87107	35
26	.42946	.90309	.44516	.89545	.46072	.88755	.47614	.87937	.49141	.87093	34
27	.42972	.90296	.44542	.89532	.46097	.88741	.47639	.87923	.49166	.87079	33
28	.42999	.90284	.44568	.89519	.46123	.88728	.47665	.87909	.49192	.87064	32
29	.43025	.90271	.44594	.89506	.46149	.88715	.47690	.87896	.49217	.87050	31
30	.43051	.90259	.44620	.89493	.46175	.88701	.47716	.87882	.49242	.87036	30
31	.43077	.90246	.44646	.89480	.46201	.88688	.47741	.87868	.49268	.87021	29
32	.43104	.90233	.44672	.89467	.46226	.88674	.47767	.87854	.49293	.87007	28
33	.43130	.90221	.44698	.89454	.46252	.88661	.47793	.87840	.49318	.86993	27
34	.43156	.90208	.44724	.89441	.46278	.88647	.47818	.87826	.49344	.86978	26
35	.43182	.90196	.44750	.89428	.46304	.88634	.47844	.87812	.49369	.86964	25
36	.43209	.90183	.44776	.89415	.46330	.88620	.47869	.87798	.49394	.86949	24
37	.43235	.90171	.44802	.89402	.46355	.88607	.47895	.87784	.49419	.86935	23
38	.43261	.90158	.44828	.89389	.46381	.88593	.47920	.87770	.49445	.86921	22
39	.43287	.90146	.44854	.89376	.46407	.88580	.47946	.87756	.49470	.86906	21
40	.43313	.90133	.44880	.89363	.46433	.88566	.47971	.87743	.49495	.86892	20
41	.43340	.90120	.44906	.89350	.46458	.88553	.47997	.87729	.49521	.86878	19
42	.43366	.90108	.44932	.89337	.46484	.88539	.48022	.87715	.49546	.86863	18
43	.43392	.90095	.44958	.89324	.46510	.88526	.48048	.87701	.49571	.86849	17
44	.43418	.90082	.44984	.89311	.46536	.88512	.48073	.87687	.49596	.86834	16
45	.43445	.90070	.45010	.89298	.46561	.88499	.48099	.87673	.49622	.86820	15
46	.43471	.90057	.45036	.89285	.46587	.88485	.48124	.87659	.49647	.86805	14
47	.43497	.90045	.45062	.89272	.46613	.88472	.48150	.87645	.49672	.86791	13
48	.43523	.90032	.45088	.89259	.46639	.88458	.48175	.87631	.49697	.86777	12
49	.43549	.90019	.45114	.89245	.46664	.88445	.48201	.87617	.49723	.86762	11
50	.43575	.90007	.45140	.89232	.46690	.88431	.48226	.87603	.49748	.86748	10
51	.43602	.89994	.45166	.89219	.46716	.88417	.48252	.87589	.49773	.86733	9
52	.43628	.89981	.45192	.89206	.46742	.88404	.48277	.87575	.49798	.86719	8
53	.43654	.89968	.45218	.89193	.46767	.88390	.48303	.87561	.49824	.86704	7
54	.43680	.89956	.45243	.89180	.46793	.88377	.48328	.87546	.49849	.86690	6
55	.43706	.89943	.45269	.89167	.46819	.88363	.48354	.87532	.49874	.86675	5
56	.43733	.89930	.45295	.89153	.46844	.88349	.48379	.87518	.49899	.86661	4
57	.43759	.89918	.45321	.89140	.46870	.88336	.48405	.87504	.49924	.86646	3
58	.43785	.89905	.45347	.89127	.46896	.88322	.48430	.87490	.49950	.86632	2
59	.43811	.89892	.45373	.89114	.46921	.88308	.48456	.87476	.49975	.86617	1
60	.43837	.89879	.45399	.89101	.46947	.88295	.48481	.87462	.50000	.86603	0
′	Cosine	Sine	Cosine	Sine	Cosine	Sine	Cosine	Sine	Cosine	Sine	′
	64°		63°		62°		61°		60°		

′	30° Sine	30° Cosine	31° Sine	31° Cosine	32° Sine	32° Cosine	33° Sine	33° Cosine	34° Sine	34° Cosine	′
0	.50000	.86603	.51504	.85717	.52992	.84805	.54464	.83867	.55919	.82904	60
1	.50025	.86588	.51529	.85702	.53017	.84789	.54488	.83851	.55943	.82887	59
2	.50050	.86573	.51554	.85687	.53041	.84774	.54513	.83835	.55968	.82871	58
3	.50076	.86559	.51579	.85672	.53066	.84759	.54537	.83819	.55992	.82855	57
4	.50101	.86544	.51604	.85657	.53091	.84743	.54561	.83804	.56016	.82839	56
5	.50126	.86530	.51628	.85642	.53115	.84728	.54586	.83788	.56040	.82822	55
6	.50151	.86515	.51653	.85627	.53140	.84712	.54610	.83772	.56064	.82806	54
7	.50176	.86501	.51678	.85612	.53164	.84697	.54635	.83756	.56088	.82790	53
8	.50201	.86486	.51703	.85597	.53189	.84681	.54659	.83740	.56112	.82773	52
9	.50227	.86471	.51728	.85582	.53214	.84666	.54683	.83724	.56136	.82757	51
10	.50252	.86457	.51753	.85567	.53238	.84650	.54708	.83708	.56160	.82741	50
11	.50277	.86442	.51778	.85551	.53263	.84635	.54732	.83692	.56184	.82724	49
12	.50302	.86427	.51803	.85536	.53288	.84619	.54756	.83676	.56208	.82708	48
13	.50327	.86413	.51828	.85521	.53312	.84604	.54781	.83660	.56232	.82692	47
14	.50352	.86398	.51852	.85506	.53337	.84588	.54805	.83645	.56256	.82675	46
15	.50377	.86384	.51877	.85491	.53361	.84573	.54829	.83629	.56280	.82659	45
16	.50403	.86369	.51902	.85476	.53386	.84557	.54854	.83613	.56305	.82643	44
17	.50428	.86354	.51927	.85461	.53411	.84542	.54878	.83597	.56329	.82626	43
18	.50453	.86340	.51952	.85446	.53435	.84526	.54902	.83581	.56353	.82610	42
19	.50478	.86325	.51977	.85431	.53460	.84511	.54927	.83565	.56377	.82593	41
20	.50503	.86310	.52002	.85416	.53484	.84495	.54951	.83549	.56401	.82577	40
21	.50528	.86295	.52026	.85401	.53509	.84480	.54975	.83533	.56425	.82561	39
22	.50553	.86281	.52051	.85385	.53534	.84464	.54999	.83517	.56449	.82544	38
23	.50578	.86266	.52076	.85370	.53558	.84448	.55024	.83501	.56473	.82528	37
24	.50603	.86251	.52101	.85355	.53583	.84433	.55048	.83485	.56497	.82511	36
25	.50628	.86237	.52126	.85340	.53607	.84417	.55072	.83469	.56521	.82495	35
26	.50654	.86222	.52151	.85325	.53632	.84402	.55097	.83453	.56545	.82478	34
27	.50679	.86207	.52175	.85310	.53656	.84386	.55121	.83437	.56569	.82462	33
28	.50704	.86192	.52200	.85294	.53681	.84370	.55145	.83421	.56593	.82446	32
29	.50729	.86178	.52225	.85279	.53705	.84355	.55169	.83405	.56617	.82429	31
30	.50754	.86163	.52250	.85264	.53730	.84339	.55194	.83389	.56641	.82413	30
31	.50779	.86148	.52275	.85249	.53754	.84324	.55218	.83373	.56665	.82396	29
32	.50804	.86133	.52299	.85234	.53779	.84308	.55242	.83356	.56689	.82380	28
33	.50829	.86119	.52324	.85218	.53804	.84292	.55266	.83340	.56713	.82363	27
34	.50854	.86104	.52349	.85203	.53828	.84277	.55291	.83324	.56736	.82347	26
35	.50879	.86089	.52374	.85188	.53853	.84261	.55315	.83308	.56760	.82330	25
36	.50904	.86074	.52399	.85173	.53877	.84245	.55339	.83292	.56784	.82314	24
37	.50929	.86059	.52423	.85157	.53902	.84230	.55363	.83276	.56808	.82297	23
38	.50954	.86045	.52448	.85142	.53926	.84214	.55388	.83260	.56832	.82281	22
39	.50979	.86030	.52473	.85127	.53951	.84198	.55412	.83244	.56856	.82264	21
40	.51004	.86015	.52498	.85112	.53975	.84182	.55436	.83228	.56880	.82248	20
41	.51029	.86000	.52522	.85096	.54000	.84167	.55460	.83212	.56904	.82231	19
42	.51054	.85985	.52547	.85081	.54024	.84151	.55484	.83195	.56928	.82214	18
43	.51079	.85970	.52572	.85066	.54049	.84135	.55509	.83179	.56952	.82198	17
44	.51104	.85956	.52597	.85051	.54073	.84120	.55533	.83163	.56976	.82181	16
45	.51129	.85941	.52621	.85035	.54097	.84104	.55557	.83147	.57000	.82165	15
46	.51154	.85926	.52646	.85020	.54122	.84088	.55581	.83131	.57024	.82148	14
47	.51179	.85911	.52671	.85005	.54146	.84072	.55605	.83115	.57047	.82132	13
48	.51204	.85896	.52696	.84989	.54171	.84057	.55630	.83098	.57071	.82115	12
49	.51229	.85881	.52720	.84974	.54195	.84041	.55654	.83082	.57095	.82098	11
50	.51254	.85866	.52745	.84959	.54220	.84025	.55678	.83066	.57119	.82082	10
51	.51279	.85851	.52770	.84943	.54244	.84009	.55702	.83050	.57143	.82065	9
52	.51304	.85836	.52794	.84928	.54269	.83994	.55726	.83034	.57167	.82048	8
53	.51329	.85821	.52819	.84913	.54293	.83978	.55750	.83017	.57191	.82032	7
54	.51354	.85806	.52844	.84897	.54317	.83962	.55775	.83001	.57215	.82015	6
55	.51379	.85792	.52869	.84882	.54342	.83946	.55799	.82985	.57238	.81999	5
56	.51404	.85777	.52893	.84866	.54366	.83930	.55823	.82969	.57262	.81982	4
57	.51429	.85762	.52918	.84851	.54391	.83915	.55847	.82953	.57286	.81965	3
58	.51454	.85747	.52943	.84836	.54415	.83899	.55871	.82936	.57310	.81949	2
59	.51479	.85732	.52967	.84820	.54440	.83883	.55895	.82920	.57334	.81932	1
60	.51504	.85717	.52992	.84805	.54464	.83867	.55919	.82904	.57358	.81915	0
′	Cosine	Sine	Cosine	Sine	Cosine	Sine	Cosine	Sine	Cosine	Sine	′
	59°		58°		57°		56°		55°		

′	35° Sine	35° Cosine	36° Sine	36° Cosine	37° Sine	37° Cosine	38° Sine	38° Cosine	39° Sine	39° Cosine	′
0	.57358	.81915	.58779	.80902	.60182	.79864	.61566	.78801	.62932	.77715	60
1	.57381	.81899	.58802	.80885	.60205	.79846	.61589	.78783	.62955	.77696	59
2	.57405	.81882	.58826	.80867	.60228	.79829	.61612	.78765	.62977	.77678	58
3	.57429	.81865	.58849	.80850	.60251	.79811	.61635	.78747	.63000	.77660	57
4	.57453	.81848	.58873	.80833	.60274	.79793	.61658	.78729	.63022	.77641	56
5	.57477	.81832	.58896	.80816	.60298	.79776	.61681	.78711	.63045	.77623	55
6	.57501	.81815	.58920	.80799	.60321	.79758	.61704	.78694	.63068	.77605	54
7	.57524	.81798	.58943	.80782	.60344	.79741	.61726	.78676	.63090	.77586	53
8	.57548	.81782	.58967	.80765	.60367	.79723	.61749	.78658	.63113	.77568	52
9	.57572	.81765	.58990	.80748	.60390	.79706	.61772	.78640	.63135	.77550	51
10	.57596	.81748	.59014	.80730	.60414	.79688	.61795	.78622	.63158	.77531	50
11	.57619	.81731	.59037	.80713	.60437	.79671	.61818	.78604	.63180	.77513	49
12	.57643	.81714	.59061	.80696	.60460	.79653	.61841	.78586	.63203	.77494	48
13	.57667	.81698	.59084	.80679	.60483	.79635	.61864	.78568	.63225	.77476	47
14	.57691	.81681	.59108	.80662	.60506	.79618	.61887	.78550	.63248	.77458	46
15	.57715	.81664	.59131	.80644	.60529	.79600	.61909	.78532	.63271	.77439	45
16	.57738	.81647	.59154	.80627	.60553	.79583	.61932	.78514	.63293	.77421	44
17	.57762	.81631	.59178	.80610	.60576	.79565	.61955	.78496	.63316	.77402	43
18	.57786	.81614	.59201	.80593	.60599	.79547	.61978	.78478	.63338	.77384	42
19	.57810	.81597	.59225	.80576	.60622	.79530	.62001	.78460	.63361	.77366	41
20	.57833	.81580	.59248	.80558	.60645	.79512	.62024	.78442	.63383	.77347	40
21	.57857	.81563	.59272	.80541	.60668	.79494	.62046	.78424	.63406	.77329	39
22	.57881	.81546	.59295	.80524	.60691	.79477	.62069	.78405	.63428	.77310	38
23	.57904	.81530	.59318	.80507	.60714	.79459	.62092	.78387	.63451	.77292	37
24	.57928	.81513	.59342	.80489	.60738	.79441	.62115	.78369	.63473	.77273	36
25	.57952	.81496	.59365	.80472	.60761	.79424	.62138	.78351	.63496	.77255	35
26	.57976	.81479	.59389	.80455	.60784	.79406	.62160	.78333	.63518	.77236	34
27	.57999	.81462	.59412	.80438	.60807	.79388	.62183	.78315	.63540	.77218	33
28	.58023	.81445	.59436	.80420	.60830	.79371	.62206	.78297	.63563	.77199	32
29	.58047	.81428	.59459	.80403	.60853	.79353	.62229	.78279	.63585	.77181	31
30	.58070	.81412	.59482	.80386	.60876	.79335	.62251	.78261	.63608	.77162	30
31	.58094	.81395	.59506	.80368	.60899	.79318	.62274	.78243	.63630	.77144	29
32	.58118	.81378	.59529	.80351	.60922	.79300	.62297	.78225	.63653	.77125	28
33	.58141	.81361	.59552	.80334	.60945	.79282	.62320	.78206	.63675	.77107	27
34	.58165	.81344	.59576	.80316	.60968	.79264	.62342	.78188	.63698	.77088	26
35	.58189	.81327	.59599	.80299	.60991	.79247	.62365	.78170	.63720	.77070	25
36	.58212	.81310	.59622	.80282	.61015	.79229	.62388	.78152	.63742	.77051	24
37	.58236	.81293	.59646	.80264	.61038	.79211	.62411	.78134	.63765	.77033	23
38	.58260	.81276	.59669	.80247	.61061	.79193	.62433	.78116	.63787	.77014	22
39	.58283	.81259	.59693	.80230	.61084	.79176	.62456	.78098	.63810	.76996	21
40	.58307	.81242	.59716	.80212	.61107	.79158	.62479	.78079	.63832	.76977	20
41	.58330	.81225	.59739	.80195	.61130	.79140	.62502	.78061	.63854	.76959	19
42	.58354	.81208	.59763	.80178	.61153	.79122	.62524	.78043	.63877	.76940	18
43	.58378	.81191	.59786	.80160	.61176	.79105	.62547	.78025	.63899	.76921	17
44	.58401	.81174	.59809	.80143	.61199	.79087	.62570	.78007	.63922	.76903	16
45	.58425	.81157	.59832	.80125	.61222	.79069	.62592	.77988	.63944	.76884	15
46	.58449	.81140	.59856	.80108	.61245	.79051	.62615	.77970	.63966	.76866	14
47	.58472	.81123	.59879	.80091	.61268	.79033	.62638	.77952	.63989	.76847	13
48	.58496	.81106	.59902	.80073	.61291	.79016	.62660	.77934	.64011	.76828	12
49	.58519	.81089	.59926	.80056	.61314	.78998	.62683	.77916	.64033	.76810	11
50	.58543	.81072	.59949	.80038	.61337	.78980	.62706	.77897	.64056	.76791	10
51	.58567	.81055	.59972	.80021	.61360	.78962	.62728	.77879	.64078	.76772	9
52	.58590	.81038	.59995	.80003	.61383	.78944	.62751	.77861	.64100	.76754	8
53	.58614	.81021	.60019	.79986	.61406	.78926	.62774	.77843	.64123	.76735	7
54	.58637	.81004	.60042	.79968	.61429	.78908	.62796	.77824	.64145	.76717	6
55	.58661	.80987	.60065	.79951	.61451	.78891	.62819	.77806	.64167	.76698	5
56	.58684	.80970	.60089	.79934	.61474	.78873	.62842	.77788	.64190	.76679	4
57	.58708	.80953	.60112	.79916	.61497	.78855	.62864	.77769	.64212	.76661	3
58	.58731	.80936	.60135	.79899	.61520	.78837	.62887	.77751	.64234	.76642	2
59	.58755	.80919	.60158	.79881	.61543	.78819	.62909	.77733	.64256	.76623	1
60	.58779	.80902	.60182	.79864	.61566	.78801	.62932	.77715	.64279	.76604	0
′	Cosine	Sine	Cosine	Sine	Cosine	Sine	Cosine	Sine	Cosine	Sine	′
	54°		53°		52°		51°		50°		

′	40° Sine	40° Cosine	41° Sine	41° Cosine	42° Sine	42° Cosine	43° Sine	43° Cosine	44° Sine	44° Cosine	′
0	.64279	.76604	.65606	.75471	.66913	.74314	.68200	.73135	.69466	.71934	60
1	.64301	.76586	.65628	.75452	.66935	.74295	.68221	.73116	.69487	.71914	59
2	.64323	.76567	.65650	.75433	.66956	.74276	.68242	.73096	.69508	.71894	58
3	.64346	.76548	.65672	.75414	.66978	.74256	.68264	.73076	.69529	.71873	57
4	.64368	.76530	.65694	.75395	.66999	.74237	.68285	.73056	.69549	.71853	56
5	.64390	.76511	.65716	.75375	.67021	.74217	.68306	.73036	.69570	.71833	55
6	.64412	.76492	.65738	.75356	.67043	.74198	.68327	.73016	.69591	.71813	54
7	.64435	.76473	.65759	.75337	.67064	.74178	.68349	.72996	.69612	.71792	53
8	.64457	.76455	.65781	.75318	.67086	.74159	.68370	.72976	.69633	.71772	52
9	.64479	.76436	.65803	.75299	.67107	.74139	.68391	.72957	.69654	.71752	51
10	.64501	.76417	.65825	.75280	.67129	.74120	.68412	.72937	.69675	.71732	50
11	.64524	.76398	.65847	.75261	.67151	.74100	.68434	.72917	.69696	.71711	49
12	.64546	.76380	.65869	.75241	.67172	.74080	.68455	.72897	.69717	.71691	48
13	.64568	.76361	.65891	.75222	.67194	.74061	.68476	.72877	.69737	.71671	47
14	.64590	.76342	.65913	.75203	.67215	.74041	.68497	.72857	.69758	.71650	46
15	.64612	.76323	.65935	.75184	.67237	.74022	.68518	.72837	.69779	.71630	45
16	.64635	.76304	.65956	.75165	.67258	.74002	.68539	.72817	.69800	.71610	44
17	.64657	.76286	.65978	.75146	.67280	.73983	.68561	.72797	.69821	.71590	43
18	.64679	.76267	.66000	.75126	.67301	.73963	.68582	.72777	.69842	.71569	42
19	.64701	.76248	.66022	.75107	.67323	.73944	.68603	.72757	.69862	.71549	41
20	.64723	.76229	.66044	.75088	.67344	.73924	.68624	.72737	.69883	.71529	40
21	.64746	.76210	.66066	.75069	.67366	.73904	.68645	.72717	.69904	.71508	39
22	.64768	.76192	.66088	.75050	.67387	.73885	.68666	.72697	.69925	.71488	38
23	.64790	.76173	.66109	.75030	.67409	.73865	.68688	.72677	.69946	.71468	37
24	.64812	.76154	.66131	.75011	.67430	.73846	.68709	.72657	.69966	.71447	36
25	.64834	.76135	.66153	.74992	.67452	.73826	.68730	.72637	.69987	.71427	35
26	.64856	.76116	.66175	.74973	.67473	.73806	.68751	.72617	.70008	.71407	34
27	.64878	.76097	.66197	.74953	.67495	.73787	.68772	.72597	.70029	.71386	33
28	.64901	.76078	.66218	.74934	.67516	.73767	.68793	.72577	.70049	.71366	32
29	.64923	.76059	.66240	.74915	.67538	.73747	.68814	.72557	.70070	.71345	31
30	.64945	.76041	.66262	.74896	.67559	.73728	.68835	.72537	.70091	.71325	30
31	.64967	.76022	.66284	.74876	.67580	.73708	.68857	.72517	.70112	.71305	29
32	.64989	.76003	.66306	.74857	.67602	.73688	.68878	.72497	.70132	.71284	28
33	.65011	.75984	.66327	.74838	.67623	.73669	.68899	.72477	.70153	.71264	27
34	.65033	.75965	.66349	.74818	.67645	.73649	.68920	.72457	.70174	.71243	26
35	.65055	.75946	.66371	.74799	.67666	.73629	.68941	.72437	.70195	.71223	25
36	.65077	.75927	.66393	.74780	.67688	.73610	.68962	.72417	.70215	.71203	24
37	.65100	.75908	.66414	.74760	.67709	.73590	.68983	.72397	.70236	.71182	23
38	.65122	.75889	.66436	.74741	.67730	.73570	.69004	.72377	.70257	.71162	22
39	.65144	.75870	.66458	.74722	.67752	.73551	.69025	.72357	.70277	.71141	21
40	.65166	.75851	.66480	.74703	.67773	.73531	.69046	.72337	.70298	.71121	20
41	.65188	.75832	.66501	.74683	.67795	.73511	.69067	.72317	.70319	.71100	19
42	.65210	.75813	.66523	.74664	.67816	.73491	.69088	.72297	.70339	.71080	18
43	.65232	.75794	.66545	.74644	.67837	.73472	.69109	.72277	.70360	.71059	17
44	.65254	.75775	.66566	.74625	.67859	.73452	.69130	.72257	.70381	.71039	16
45	.65276	.75756	.66588	.74606	.67880	.73432	.69151	.72236	.70401	.71019	15
46	.65298	.75738	.66610	.74586	.67901	.73413	.69172	.72216	.70422	.70998	14
47	.65320	.75719	.66632	.74567	.67923	.73393	.69193	.72196	.70443	.70978	13
48	.65342	.75700	.66653	.74548	.67944	.73373	.69214	.72176	.70463	.70957	12
49	.65364	.75680	.66675	.74528	.67965	.73353	.69235	.72156	.70484	.70937	11
50	.65386	.75661	.66697	.74509	.67987	.73333	.69256	.72136	.70505	.70916	10
51	.65408	.75642	.66718	.74489	.68008	.73314	.69277	.72116	.70525	.70896	9
52	.65430	.75623	.66740	.74470	.68029	.73294	.69298	.72095	.70546	.70875	8
53	.65452	.75604	.66762	.74451	.68051	.73274	.69319	.72075	.70567	.70855	7
54	.65474	.75585	.66783	.74431	.68072	.73254	.69340	.72055	.70587	.70834	6
55	.65496	.75566	.66805	.74412	.68093	.73234	.69361	.72035	.70608	.70813	5
56	.65518	.75547	.66827	.74392	.68115	.73215	.69382	.72015	.70628	.70793	4
57	.65540	.75528	.66848	.74373	.68136	.73195	.69403	.71995	.70649	.70772	3
58	.65562	.75509	.66870	.74353	.68157	.73175	.69424	.71974	.70670	.70752	2
59	.65584	.75490	.66891	.74334	.68179	.73155	.69445	.71954	.70690	.70731	1
60	.65606	.75471	.66913	.74314	.68200	.73135	.69466	.71934	.70711	.70711	0
′	Cosine	Sine	Cosine	Sine	Cosine	Sine	Cosine	Sine	Cosine	Sine	′

| 49° | 48° | 47° | 46° | 45° |

′	0° Tang	Cotang	1° Tang	Cotang	2° Tang	Cotang	3° Tang	Cotang	4° Tang	Cotang	′
0	.00000	Infin.	.01746	57.2900	.03492	28.6363	.05241	19.0811	.06993	14.3007	60
1	.00029	3437.75	.01775	56.3506	.03521	28.3994	.05270	18.9755	.07022	14.2411	59
2	.00058	1718.87	.01804	55.4415	.03550	28.1664	.05299	18.8711	.07051	14.1821	58
3	.00087	1145.92	.01833	54.5613	.03579	27.9372	.05328	18.7678	.07080	14.1235	57
4	.00116	859.436	.01862	53.7086	.03609	27.7117	.05357	18.6656	.07110	14.0655	56
5	.00145	687.549	.01891	52.8821	.03638	27.4899	.05387	18.5645	.07139	14.0079	55
6	.00175	572.957	.01920	52.0807	.03667	27.2715	.05416	18.4645	.07168	13.9507	54
7	.00204	491.106	.01949	51.3032	.03696	27.0566	.05445	18.3655	.07197	13.8940	53
8	.00233	429.718	.01978	50.5485	.03725	26.8450	.05474	18.2677	.07227	13.8378	52
9	.00262	381.971	.02007	49.8157	.03754	26.6367	.05503	18.1708	.07256	13.7821	51
10	.00291	343.774	.02036	49.1039	.03783	26.4316	.05533	18.0750	.07285	13.7267	50
11	.00320	312.521	.02066	48.4121	.03812	26.2296	.05562	17.9802	.07314	13.6719	49
12	.00349	286.478	.02095	47.7395	.03842	26.0307	.05591	17.8863	.07344	13.6174	48
13	.00378	264.441	.02124	47.0853	.03871	25.8348	.05620	17.7934	.07373	13.5634	47
14	.00407	245.552	.02153	46.4489	.03900	25.6418	.05649	17.7015	.07402	13.5098	46
15	.00436	229.182	.02182	45.8294	.03929	25.4517	.05678	17.6106	.07431	13.4566	45
16	.00465	214.858	.02211	45.2261	.03958	25.2644	.05708	17.5205	.07461	13.4039	44
17	.00495	202.219	.02240	44.6386	.03987	25.0798	.05737	17.4314	.07490	13.3515	43
18	.00524	190.984	.02269	44.0661	.04016	24.8978	.05766	17.3432	.07519	13.2996	42
19	.00553	180.932	.02298	43.5081	.04046	24.7185	.05795	17.2558	.07548	13.2480	41
20	.00582	171.885	.02328	42.9641	.04075	24.5418	.05824	17.1693	.07578	13.1969	40
21	.00611	163.700	.02357	42.4335	.04104	24.3675	.05854	17.0837	.07607	13.1461	39
22	.00640	156.259	.02386	41.9158	.04133	24.1957	.05883	16.9990	.07636	13.0958	38
23	.00669	149.465	.02415	41.4106	.04162	24.0263	.05912	16.9150	.07665	13.0458	37
24	.00698	143.237	.02444	40.9174	.04191	23.8593	.05941	16.8319	.07695	12.9962	36
25	.00727	137.507	.02473	40.4358	.04220	23.6945	.05970	16.7496	.07724	12.9469	35
26	.00756	132.219	.02502	39.9655	.04250	23.5321	.05999	16.6681	.07753	12.8981	34
27	.00785	127.321	.02531	39.5059	.04279	23.3718	.06029	16.5874	.07782	12.8496	33
28	.00815	122.774	.02560	39.0568	.04308	23.2137	.06058	16.5075	.07812	12.8014	32
29	.00844	118.540	.02589	38.6177	.04337	23.0577	.06087	16.4283	.07841	12.7536	31
30	.00873	114.589	.02619	38.1885	.04366	22.9038	.06116	16.3499	.07870	12.7062	30
31	.00902	110.892	.02648	37.7686	.04395	22.7519	.06145	16.2722	.07899	12.6591	29
32	.00931	107.426	.02677	37.3579	.04424	22.6020	.06175	16.1952	.07929	12.6124	28
33	.00960	104.171	.02706	36.9560	.04454	22.4541	.06204	16.1190	.07958	12.5660	27
34	.00989	101.107	.02735	36.5627	.04483	22.3081	.06233	16.0435	.07987	12.5199	26
35	.01018	98.2179	.02764	36.1776	.04512	22.1640	.06262	15.9687	.08017	12.4742	25
36	.01047	95.4895	.02793	35.8006	.04541	22.0217	.06291	15.8945	.08046	12.4288	24
37	.01076	92.9085	.02822	35.4313	.04570	21.8813	.06321	15.8211	.08075	12.3838	23
38	.01105	90.4633	.02851	35.0695	.04599	21.7426	.06350	15.7483	.08104	12.3390	22
39	.01135	88.1436	.02881	34.7151	.04628	21.6056	.06379	15.6762	.08134	12.2946	21
40	.01164	85.9398	.02910	34.3678	.04658	21.4704	.06408	15.6048	.08163	12.2505	20
41	.01193	83.8435	.02939	34.0273	.04687	21.3369	.06437	15.5340	.08192	12.2067	19
42	.01222	81.8470	.02968	33.6935	.04716	21.2049	.06467	15.4638	.08221	12.1632	18
43	.01251	79.9434	.02997	33.3662	.04745	21.0747	.06496	15.3943	.08251	12.1201	17
44	.01280	78.1263	.03026	33.0452	.04774	20.9460	.06525	15.3254	.08280	12.0772	16
45	.01309	76.3900	.03055	32.7303	.04803	20.8188	.06554	15.2571	.08309	12.0346	15
46	.01338	74.7292	.03084	32.4213	.04833	20.6932	.06584	15.1893	.08339	11.9923	14
47	.01367	73.1390	.03114	32.1181	.04862	20.5691	.06613	15.1222	.08368	11.9504	13
48	.01396	71.6151	.03143	31.8205	.04891	20.4465	.06642	15.0557	.08397	11.9087	12
49	.01425	70.1533	.03172	31.5284	.04920	20.3253	.06671	14.9898	.08427	11.8673	11
50	.01455	68.7501	.03201	31.2416	.04949	20.2056	.06700	14.9244	.08456	11.8262	10
51	.01484	67.4019	.03230	30.9599	.04978	20.0872	.06730	14.8596	.08485	11.7853	9
52	.01513	66.1055	.03259	30.6833	.05007	19.9702	.06759	14.7954	.08514	11.7448	8
53	.01542	64.8580	.03288	30.4116	.05037	19.8546	.06788	14.7317	.08544	11.7045	7
54	.01571	63.6567	.03317	30.1446	.05066	19.7403	.06817	14.6685	.08573	11.6645	6
55	.01600	62.4992	.03346	29.8823	.05095	19.6273	.06847	14.6059	.08602	11.6248	5
56	.01629	61.3829	.03376	29.6245	.05124	19.5156	.06876	14.5438	.08632	11.5853	4
57	.01658	60.3058	.03405	29.3711	.05153	19.4051	.06905	14.4823	.08661	11.5461	3
58	.01687	59.2659	.03434	29.1220	.05182	19.2959	.06934	14.4212	.08690	11.5072	2
59	.01716	58.2612	.03463	28.8771	.05212	19.1879	.06963	14.3607	.08720	11.4685	1
60	.01746	57.2900	.03492	28.6363	.05241	19.0811	.06993	14.3007	.08749	11.4301	0
′	Cotang	Tang	Cotang	Tang	Cotang	Tang	Cotang	Tang	Cotang	Tang	′

| 89° | 88° | 87° | 86° | 85° |

′	5° Tang	Cotang	6° Tang	Cotang	7° Tang	Cotang	8° Tang	Cotang	9° Tang	Cotang	′
0	.08749	11.4301	.10510	9.51436	.12278	8.14435	.14054	7.11537	.15838	6.31375	60
1	.08778	11.3919	.10540	9.48781	.12308	8.12481	.14084	7.10038	.15868	6.30189	59
2	.08807	11.3540	.10569	9.46141	.12338	8.10536	.14113	7.08546	.15898	6.29007	58
3	.08837	11.3163	.10599	9.43515	.12367	8.08600	.14143	7.07059	.15928	6.27829	57
4	.08866	11.2789	.10628	9.40904	.12397	8.06674	.14173	7.05579	.15958	6.26655	56
5	.08895	11.2417	.10657	9.38307	.12426	8.04756	.14202	7.04105	.15988	6.25486	55
6	.08925	11.2048	.10687	9.35724	.12456	8.02848	.14232	7.02637	.16017	6.24321	54
7	.08954	11.1681	.10716	9.33155	.12485	8.00948	.14262	7.01174	.16047	6.23160	53
8	.08983	11.1316	.10746	9.30599	.12515	7.99058	.14291	6.99718	.16077	6.22003	52
9	.09013	11.0954	.10775	9.28058	.12544	7.97176	.14321	6.98268	.16107	6.20851	51
10	.09042	11.0594	.10805	9.25530	.12574	7.95302	.14351	6.96823	.16137	6.19703	50
11	.09071	11.0237	.10834	9.23016	.12603	7.93438	.14381	6.95385	.16167	6.18559	49
12	.09101	10.9882	.10863	9.20516	.12633	7.91582	.14410	6.93952	.16196	6.17419	48
13	.09130	10.9529	.10893	9.18028	.12662	7.89734	.14440	6.92525	.16226	6.16283	47
14	.09159	10.9178	.10922	9.15554	.12692	7.87895	.14470	6.91104	.16256	6.15151	46
15	.09189	10.8829	.10952	9.13093	.12722	7.86064	.14499	6.89688	.16286	6.14023	45
16	.09218	10.8483	.10981	9.10646	.12751	7.84242	.14529	6.88278	.16316	6.12899	44
17	.09247	10.8139	.11011	9.08211	.12781	7.82428	.14559	6.86874	.16346	6.11779	43
18	.09277	10.7797	.11040	9.05789	.12810	7.80622	.14588	6.85475	.16376	6.10664	42
19	.09306	10.7457	.11070	9.03379	.12840	7.78825	.14618	6.84082	.16405	6.09552	41
20	.09335	10.7119	.11099	9.00983	.12869	7.77035	.14648	6.82694	.16435	6.08444	40
21	.09365	10.6783	.11128	8.98598	.12899	7.75254	.14678	6.81312	.16465	6.07340	39
22	.09394	10.6450	.11158	8.96227	.12929	7.73480	.14707	6.79936	.16495	6.06240	38
23	.09423	10.6118	.11187	8.93867	.12958	7.71715	.14737	6.78564	.16525	6.05143	37
24	.09453	10.5789	.11217	8.91520	.12988	7.69957	.14767	6.77199	.16555	6.04051	36
25	.09482	10.5462	.11246	8.89185	.13017	7.68208	.14796	6.75838	.16585	6.02962	35
26	.09511	10.5136	.11276	8.86862	.13047	7.66466	.14826	6.74483	.16615	6.01878	34
27	.09541	10.4813	.11305	8.84551	.13076	7.64732	.14856	6.73133	.16645	6.00797	33
28	.09570	10.4491	.11335	8.82252	.13106	7.63005	.14886	6.71789	.16674	5.99720	32
29	.09600	10.4172	.11364	8.79964	.13136	7.61287	.14915	6.70450	.16704	5.98646	31
30	.09629	10.3854	.11394	8.77689	.13165	7.59575	.14945	6.69116	.16734	5.97576	30
31	.09658	10.3538	.11423	8.75425	.13195	7.57872	.14975	6.67787	.16764	5.96510	29
32	.09688	10.3224	.11452	8.73172	.13224	7.56176	.15005	6.66463	.16794	5.95448	28
33	.09717	10.2913	.11482	8.70931	.13254	7.54487	.15034	6.65144	.16824	5.94390	27
34	.09746	10.2602	.11511	8.68701	.13284	7.52806	.15064	6.63831	.16854	5.93335	26
35	.09776	10.2294	.11541	8.66482	.13313	7.51132	.15094	6.62523	.16884	5.92283	25
36	.09805	10.1988	.11570	8.64275	.13343	7.49465	.15124	6.61219	.16914	5.91236	24
37	.09834	10.1683	.11600	8.62078	.13372	7.47806	.15153	6.59921	.16944	5.90191	23
38	.09864	10.1381	.11629	8.59893	.13402	7.46154	.15183	6.58627	.16974	5.89151	22
39	.09893	10.1080	.11659	8.57718	.13432	7.44509	.15213	6.57339	.17004	5.88114	21
40	.09923	10.0780	.11688	8.55555	.13461	7.42871	.15243	6.56055	.17033	5.87080	20
41	.09952	10.0483	.11718	8.53402	.13491	7.41240	.15272	6.54777	.17063	5.86051	19
42	.09981	10.0187	.11747	8.51259	.13521	7.39616	.15302	6.53503	.17093	5.85024	18
43	.10011	9.98931	.11777	8.49128	.13550	7.37999	.15332	6.52234	.17123	5.84001	17
44	.10040	9.96007	.11806	8.47007	.13580	7.36389	.15362	6.50970	.17153	5.82982	16
45	.10069	9.93101	.11836	8.44896	.13609	7.34786	.15391	6.49710	.17183	5.81966	15
46	.10099	9.90211	.11865	8.42795	.13639	7.33190	.15421	6.48456	.17213	5.80953	14
47	.10128	9.87338	.11895	8.40705	.13669	7.31600	.15451	6.47206	.17243	5.79944	13
48	.10158	9.84482	.11924	8.38625	.13698	7.30018	.15481	6.45961	.17273	5.78938	12
49	.10187	9.81641	.11954	8.36555	.13728	7.28442	.15511	6.44720	.17303	5.77936	11
50	.10216	9.78817	.11983	8.34496	.13758	7.26873	.15540	6.43484	.17333	5.76937	10
51	.10246	9.76009	.12013	8.32446	.13787	7.25310	.15570	6.42253	.17363	5.75941	9
52	.10275	9.73217	.12042	8.30406	.13817	7.23754	.15600	6.41026	.17393	5.74949	8
53	.10305	9.70441	.12072	8.28376	.13846	7.22204	.15630	6.39804	.17423	5.73960	7
54	.10334	9.67680	.12101	8.26355	.13876	7.20661	.15660	6.38587	.17453	5.72974	6
55	.10363	9.64935	.12131	8.24345	.13906	7.19125	.15689	6.37374	.17483	5.71992	5
56	.10393	9.62205	.12160	8.22344	.13935	7.17594	.15719	6.36165	.17513	5.71013	4
57	.10422	9.59490	.12190	8.20352	.13965	7.16071	.15749	6.34961	.17543	5.70037	3
58	.10452	9.56791	.12219	8.18370	.13995	7.14553	.15779	6.33761	.17573	5.69064	2
59	.10481	9.54106	.12249	8.16398	.14024	7.13042	.15809	6.32566	.17603	5.68094	1
60	.10510	9.51436	.12278	8.14435	.14054	7.11537	.15838	6.31375	.17633	5.67128	0
′	Cotang	Tang	Cotang	Tang	Cotang	Tang	Cotang	Tang	Cotang	Tang	′
	84°		83°		82°		81°		80°		

′	10° Tang	Cotang	11° Tang	Cotang	12° Tang	Cotang	13° Tang	Cotang	14° Tang	Cotang	′
0	.17633	5.67128	.19438	5.14455	.21256	4.70463	.23087	4.33148	.24933	4.01078	60
1	.17663	5.66165	.19468	5.13658	.21286	4.69791	.23117	4.32573	.24964	4.00582	59
2	.17693	5.65205	.19498	5.12862	.21316	4.69121	.23148	4.32001	.24995	4.00086	58
3	.17723	5.64248	.19529	5.12069	.21347	4.68452	.23179	4.31430	.25026	3.99592	57
4	.17753	5.63295	.19559	5.11279	.21377	4.67786	.23209	4.30860	.25056	3.99099	56
5	.17783	5.62344	.19589	5.10490	.21408	4.67121	.23240	4.30291	.25087	3.98607	55
6	.17813	5.61397	.19619	5.09704	.21438	4.66458	.23271	4.29724	.25118	3.98117	54
7	.17843	5.60452	.19649	5.08921	.21469	4.65797	.23301	4.29159	.25149	3.97627	53
8	.17873	5.59511	.19680	5.08139	.21499	4.65138	.23332	4.28595	.25180	3.97139	52
9	.17903	5.58573	.19710	5.07360	.21529	4.64480	.23363	4.28032	.25211	3.96651	51
10	.17933	5.57638	.19740	5.06584	.21560	4.63825	.23393	4.27471	.25242	3.96165	50
11	.17963	5.56706	.19770	5.05809	.21590	4.63171	.23424	4.26911	.25273	3.95680	49
12	.17993	5.55777	.19801	5.05037	.21621	4.62518	.23455	4.26352	.25304	3.95196	48
13	.18023	5.54851	.19831	5.04267	.21651	4.61868	.23485	4.25795	.25335	3.94713	47
14	.18053	5.53927	.19861	5.03499	.21682	4.61219	.23516	4.25239	.25366	3.94232	46
15	.18083	5.53007	.19891	5.02734	.21712	4.60572	.23547	4.24685	.25397	3.93751	45
16	.18113	5.52090	.19921	5.01971	.21743	4.59927	.23578	4.24132	.25428	3.93271	44
17	.18143	5.51176	.19952	5.01210	.21773	4.59283	.23608	4.23580	.25459	3.92793	43
18	.18173	5.50264	.19982	5.00451	.21804	4.58641	.23639	4.23030	.25490	3.92316	42
19	.18203	5.49356	.20012	4.99695	.21834	4.58001	.23670	4.22481	.25521	3.91839	41
20	.18233	5.48451	.20042	4.98940	.21864	4.57363	.23700	4.21933	.25552	3.91364	40
21	.18263	5.47548	.20073	4.98188	.21895	4.56726	.23731	4.21387	.25583	3.90890	39
22	.18293	5.46648	.20103	4.97438	.21925	4.56091	.23762	4.20842	.25614	3.90417	38
23	.18323	5.45751	.20133	4.96690	.21956	4.55458	.23793	4.20298	.25645	3.89945	37
24	.18353	5.44857	.20164	4.95945	.21986	4.54826	.23823	4.19756	.25676	3.89474	36
25	.18384	5.43966	.20194	4.95201	.22017	4.54196	.23854	4.19215	.25707	3.89004	35
26	.18414	5.43077	.20224	4.94460	.22047	4.53568	.23885	4.18675	.25738	3.88536	34
27	.18444	5.42192	.20254	4.93721	.22078	4.52941	.23916	4.18137	.25769	3.88068	33
28	.18474	5.41309	.20285	4.92984	.22108	4.52316	.23946	4.17600	.25800	3.87601	32
29	.18504	5.40429	.20315	4.92249	.22139	4.51693	.23977	4.17064	.25831	3.87136	31
30	.18534	5.39552	.20345	4.91516	.22169	4.51071	.24008	4.16530	.25862	3.86671	30
31	.18564	5.38677	.20376	4.90785	.22200	4.50451	.24039	4.15997	.25893	3.86208	29
32	.18594	5.37805	.20406	4.90056	.22231	4.49832	.24069	4.15465	.25924	3.85745	28
33	.18624	5.36936	.20436	4.89330	.22261	4.49215	.24100	4.14934	.25955	3.85284	27
34	.18654	5.36070	.20466	4.88605	.22292	4.48600	.24131	4.14405	.25986	3.84824	26
35	.18684	5.35206	.20497	4.87882	.22322	4.47986	.24162	4.13877	.26017	3.84364	25
36	.18714	5.34345	.20527	4.87162	.22353	4.47374	.24193	4.13350	.26048	3.83906	24
37	.18745	5.33487	.20557	4.86444	.22383	4.46764	.24223	4.12825	.26079	3.83449	23
38	.18775	5.32631	.20588	4.85727	.22414	4.46155	.24254	4.12301	.26110	3.82992	22
39	.18805	5.31778	.20618	4.85013	.22444	4.45548	.24285	4.11778	.26141	3.82537	21
40	.18835	5.30928	.20648	4.84300	.22475	4.44942	.24316	4.11256	.26172	3.82083	20
41	.18865	5.30080	.20679	4.83590	.22505	4.44338	.24347	4.10736	.26203	3.81630	19
42	.18895	5.29235	.20709	4.82882	.22536	4.43735	.24377	4.10216	.26235	3.81177	18
43	.18925	5.28393	.20739	4.82175	.22567	4.43134	.24408	4.09699	.26266	3.80726	17
44	.18955	5.27553	.20770	4.81471	.22597	4.42534	.24439	4.09182	.26297	3.80276	16
45	.18986	5.26715	.20800	4.80769	.22628	4.41936	.24470	4.08666	.26328	3.79827	15
46	.19016	5.25880	.20830	4.80068	.22658	4.41340	.24501	4.08152	.26359	3.79378	14
47	.19046	5.25048	.20861	4.79370	.22689	4.40745	.24532	4.07639	.26390	3.78931	13
48	.19076	5.24218	.20891	4.78673	.22719	4.40152	.24562	4.07127	.26421	3.78485	12
49	.19106	5.23391	.20921	4.77978	.22750	4.39560	.24593	4.06616	.26452	3.78040	11
50	.19136	5.22566	.20952	4.77286	.22781	4.38969	.24624	4.06107	.26483	3.77595	10
51	.19166	5.21744	.20982	4.76595	.22811	4.38381	.24655	4.05599	.26515	3.77152	9
52	.19197	5.20925	.21013	4.75906	.22842	4.37793	.24686	4.05092	.26546	3.76709	8
53	.19227	5.20107	.21043	4.75219	.22872	4.37207	.24717	4.04586	.26577	3.76268	7
54	.19257	5.19293	.21073	4.74534	.22903	4.36623	.24747	4.04081	.26608	3.75828	6
55	.19287	5.18480	.21104	4.73851	.22934	4.36040	.24778	4.03578	.26639	3.75388	5
56	.19317	5.17671	.21134	4.73170	.22964	4.35459	.24809	4.03076	.26670	3.74950	4
57	.19347	5.16863	.21164	4.72490	.22995	4.34879	.24840	4.02574	.26701	3.74512	3
58	.19378	5.16058	.21195	4.71813	.23026	4.34300	.24871	4.02074	.26733	3.74075	2
59	.19408	5.15256	.21225	4.71137	.23056	4.33723	.24902	4.01576	.26764	3.73640	1
60	.19438	5.14455	.21256	4.70463	.23087	4.33148	.24933	4.01078	.26795	3.73205	0
′	Cotang	Tang	Cotang	Tang	Cotang	Tang	Cotang	Tang	Cotang	Tang	′
	79°		78°		77°		76°		75°		

′	15° Tang	Cotang	16° Tang	Cotang	17° Tang	Cotang	18° Tang	Cotang	19° Tang	Cotang	′
0	.26795	3.73205	.28675	3.48741	.30573	3.27085	.32492	3.07768	.34433	2.90421	60
1	.26826	3.72771	.28706	3.48359	.30605	3.26745	.32524	3.07464	.34465	2.90147	59
2	.26857	3.72338	.28738	3.47977	.30637	3.26406	.32556	3.07160	.34498	2.89873	58
3	.26888	3.71907	.28769	3.47596	.30669	3.26067	.32588	3.06857	.34530	2.89600	57
4	.26920	3.71476	.28800	3.47216	.30700	3.25729	.32621	3.06554	.34563	2.89327	56
5	.26951	3.71046	.28832	3.46837	.30732	3.25392	.32653	3.06252	.34596	2.89055	55
6	.26982	3.70616	.28864	3.46458	.30764	3.25055	.32685	3.05950	.34628	2.88783	54
7	.27013	3.70188	.28895	3.46080	.30796	3.24719	.32717	3.05649	.34661	2.88511	53
8	.27044	3.69761	.28927	3.45703	.30828	3.24383	.32749	3.05349	.34693	2.88240	52
9	.27076	3.69335	.28958	3.45327	.30860	3.24049	.32782	3.05049	.34726	2.87970	51
10	.27107	3.68909	.28990	3.44951	.30891	3.23714	.32814	3.04749	.34758	2.87700	50
11	.27138	3.68485	.29021	3.44576	.30923	3.23381	.32846	3.04450	.34791	2.87430	49
12	.27169	3.68061	.29053	3.44202	.30955	3.23048	.32878	3.04152	.34824	2.87161	48
13	.27201	3.67638	.29084	3.43829	.30987	3.22715	.32911	3.03854	.34856	2.86892	47
14	.27232	3.67217	.29116	3.43456	.31019	3.22384	.32943	3.03556	.34889	2.86624	46
15	.27263	3.66795	.29147	3.43084	.31051	3.22053	.32975	3.03260	.34922	2.86356	45
16	.27294	3.66376	.29179	3.42713	.31083	3.21722	.33007	3.02963	.34954	2.86089	44
17	.27326	3.65957	.29210	3.42343	.31115	3.21392	.33040	3.02667	.34987	2.85822	43
18	.27357	3.65538	.29242	3.41973	.31147	3.21063	.33072	3.02372	.35020	2.85555	42
19	.27388	3.65121	.29274	3.41604	.31178	3.20734	.33104	3.02077	.35052	2.85289	41
20	.27419	3.64705	.29305	3.41236	.31210	3.20406	.33136	3.01783	.35085	2.85023	40
21	.27451	3.64289	.29337	3.40869	.31242	3.20079	.33169	3.01489	.35118	2.84758	39
22	.27482	3.63874	.29368	3.40502	.31274	3.19752	.33201	3.01196	.35150	2.84494	38
23	.27513	3.63461	.29400	3.40136	.31306	3.19426	.33233	3.00903	.35183	2.84229	37
24	.27545	3.63048	.29432	3.39771	.31338	3.19100	.33266	3.00611	.35216	2.83965	36
25	.27576	3.62636	.29463	3.39406	.31370	3.18775	.33298	3.00319	.35248	2.83702	35
26	.27607	3.62224	.29495	3.39042	.31402	3.18451	.33330	3.00028	.35281	2.83439	34
27	.27638	3.61814	.29526	3.38679	.31434	3.18127	.33363	2.99738	.35314	2.83176	33
28	.27670	3.61405	.29558	3.38317	.31466	3.17804	.33395	2.99447	.35346	2.82914	32
29	.27701	3.60996	.29590	3.37955	.31498	3.17481	.33427	2.99158	.35379	2.82653	31
30	.27732	3.60588	.29621	3.37594	.31530	3.17159	.33460	2.98868	.35412	2.82391	30
31	.27764	3.60181	.29653	3.37234	.31562	3.16838	.33492	2.98580	.35445	2.82130	29
32	.27795	3.59775	.29685	3.36875	.31594	3.16517	.33524	2.98292	.35477	2.81870	28
33	.27826	3.59370	.29716	3.36516	.31626	3.16197	.33557	2.98004	.35510	2.81610	27
34	.27858	3.58966	.29748	3.36158	.31658	3.15877	.33589	2.97717	.35543	2.81350	26
35	.27889	3.58562	.29780	3.35800	.31690	3.15558	.33621	2.97430	.35576	2.81091	25
36	.27921	3.58160	.29811	3.35443	.31722	3.15240	.33654	2.97144	.35608	2.80833	24
37	.27952	3.57758	.29843	3.35087	.31754	3.14922	.33686	2.96858	.35641	2.80574	23
38	.27983	3.57357	.29875	3.34732	.31786	3.14605	.33718	2.96573	.35674	2.80316	22
39	.28015	3.56957	.29906	3.34377	.31818	3.14288	.33751	2.96288	.35707	2.80059	21
40	.28046	3.56557	.29938	3.34023	.31850	3.13972	.33783	2.96004	.35740	2.79802	20
41	.28077	3.56159	.29970	3.33670	.31882	3.13656	.33816	2.95721	.35772	2.79545	19
42	.28109	3.55761	.30001	3.33317	.31914	3.13341	.33848	2.95437	.35805	2.79289	18
43	.28140	3.55364	.30033	3.32965	.31946	3.13027	.33881	2.95155	.35838	2.79033	17
44	.28172	3.54968	.30065	3.32614	.31978	3.12713	.33913	2.94872	.35871	2.78778	16
45	.28203	3.54573	.30097	3.32264	.32010	3.12400	.33945	2.94591	.35904	2.78523	15
46	.28234	3.54179	.30128	3.31914	.32042	3.12087	.33978	2.94309	.35937	2.78269	14
47	.28266	3.53785	.30160	3.31565	.32074	3.11775	.34010	2.94028	.35969	2.78014	13
48	.28297	3.53393	.30192	3.31216	.32106	3.11464	.34043	2.93748	.36002	2.77761	12
49	.28329	3.53001	.30224	3.30868	.32139	3.11153	.34075	2.93468	.36035	2.77507	11
50	.28360	3.52609	.30255	3.30521	.32171	3.10842	.34108	2.93189	.36068	2.77254	10
51	.28391	3.52219	.30287	3.30174	.32203	3.10532	.34140	2.92910	.36101	2.77002	9
52	.28423	3.51829	.30319	3.29828	.32235	3.10223	.34173	2.92632	.36134	2.76750	8
53	.28454	3.51441	.30351	3.29483	.32267	3.09914	.34205	2.92354	.36167	2.76498	7
54	.28486	3.51053	.30382	3.29139	.32299	3.09606	.34238	2.92076	.36199	2.76247	6
55	.28517	3.50666	.30414	3.28795	.32331	3.09298	.34270	2.91799	.36232	2.75996	5
56	.28549	3.50279	.30446	3.28452	.32363	3.08991	.34303	2.91523	.36265	2.75746	4
57	.28580	3.49894	.30478	3.28109	.32396	3.08685	.34335	2.91246	.36298	2.75496	3
58	.28612	3.49509	.30509	3.27767	.32428	3.08379	.34368	2.90971	.36331	2.75246	2
59	.28643	3.49125	.30541	3.27426	.32460	3.08073	.34400	2.90696	.36364	2.74997	1
60	.28675	3.48741	.30573	3.27085	.32492	3.07768	.34433	2.90421	.36397	2.74748	0
′	Cotang	Tang	Cotang	Tang	Cotang	Tang	Cotang	Tang	Cotang	Tang	′
	74°		73°		72°		71°		70°		

NATURAL TANGENTS AND COTANGENTS.

′	20° Tang	Cotang	21° Tang	Cotang	22° Tang	Cotang	23° Tang	Cotang	24° Tang	Cotang	′
0	.36397	2.74748	.38386	2.60509	.40403	2.47509	.42447	2.35585	.44523	2.24604	60
1	.36430	2.74499	.38420	2.60283	.40436	2.47302	.42482	2.35395	.44558	2.24428	59
2	.36463	2.74251	.38453	2.60057	.40470	2.47095	.42516	2.35205	.44593	2.24252	58
3	.36496	2.74004	.38487	2.59831	.40504	2.46888	.42551	2.35015	.44627	2.24077	57
4	.36529	2.73756	.38520	2.59606	.40538	2.46682	.42585	2.34825	.44662	2.23902	56
5	.36562	2.73509	.38553	2.59381	.40572	2.46476	.42619	2.34636	.44697	2.23727	55
6	.36595	2.73263	.38587	2.59156	.40606	2.46270	.42654	2.34447	.44732	2.23553	54
7	.36628	2.73017	.38620	2.58932	.40640	2.46065	.42688	2.34258	.44767	2.23378	53
8	.36661	2.72771	.38654	2.58708	.40674	2.45860	.42722	2.34069	.44802	2.23204	52
9	.36694	2.72526	.38687	2.58484	.40707	2.45655	.42757	2.33881	.44837	2.23030	51
10	.36727	2.72281	.38721	2.58261	.40741	2.45451	.42791	2.33693	.44872	2.22857	50
11	.36760	2.72036	.38754	2.58038	.40775	2.45246	.42826	2.33505	.44907	2.22683	49
12	.36793	2.71792	.38787	2.57815	.40809	2.45043	.42860	2.33317	.44942	2.22510	48
13	.36826	2.71548	.38821	2.57593	.40843	2.44839	.42894	2.33130	.44977	2.22337	47
14	.36859	2.71305	.38854	2.57371	.40877	2.44636	.42929	2.32943	.45012	2.22164	46
15	.36892	2.71062	.38888	2.57150	.40911	2.44433	.42963	2.32756	.45047	2.21992	45
16	.36925	2.70819	.38921	2.56928	.40945	2.44230	.42998	2.32570	.45082	2.21819	44
17	.36958	2.70577	.38955	2.56707	.40979	2.44027	.43032	2.32383	.45117	2.21647	43
18	.36991	2.70335	.38988	2.56487	.41013	2.43825	.43067	2.32197	.45152	2.21475	42
19	.37024	2.70094	.39022	2.56266	.41047	2.43623	.43101	2.32012	.45187	2.21304	41
20	.37057	2.69853	.39055	2.56046	.41081	2.43422	.43136	2.31826	.45222	2.21132	40
21	.37090	2.69612	.39089	2.55827	.41115	2.43220	.43170	2.31641	.45257	2.20961	39
22	.37123	2.69371	.39122	2.55608	.41149	2.43019	.43205	2.31456	.45292	2.20790	38
23	.37157	2.69131	.39156	2.55389	.41183	2.42819	.43239	2.31271	.45327	2.20619	37
24	.37190	2.68892	.39190	2.55170	.41217	2.42618	.43274	2.31086	.45362	2.20449	36
25	.37223	2.68653	.39223	2.54952	.41251	2.42418	.43308	2.30902	.45397	2.20278	35
26	.37256	2.68414	.39257	2.54734	.41285	2.42218	.43343	2.30718	.45432	2.20108	34
27	.37289	2.68175	.39290	2.54516	.41319	2.42019	.43378	2.30534	.45467	2.19938	33
28	.37322	2.67937	.39324	2.54299	.41353	2.41819	.43412	2.30351	.45502	2.19769	32
29	.37355	2.67700	.39357	2.54082	.41387	2.41620	.43447	2.30167	.45538	2.19599	31
30	.37388	2.67462	.39391	2.53865	.41421	2.41421	.43481	2.29984	.45573	2.19430	30
31	.37422	2.67225	.39425	2.53648	.41455	2.41223	.43516	2.29801	.45608	2.19261	29
32	.37455	2.66989	.39458	2.53432	.41490	2.41025	.43550	2.29619	.45643	2.19092	28
33	.37488	2.66752	.39492	2.53217	.41524	2.40827	.43585	2.29437	.45678	2.18923	27
34	.37521	2.66516	.39526	2.53001	.41558	2.40629	.43620	2.29254	.45713	2.18755	26
35	.37554	2.66281	.39559	2.52786	.41592	2.40432	.43654	2.29073	.45748	2.18587	25
36	.37588	2.66046	.39593	2.52571	.41626	2.40235	.43689	2.28891	.45784	2.18419	24
37	.37621	2.65811	.39626	2.52357	.41660	2.40038	.43724	2.28710	.45819	2.18251	23
38	.37654	2.65576	.39660	2.52142	.41694	2.39841	.43758	2.28528	.45854	2.18084	22
39	.37687	2.65342	.39694	2.51929	.41728	2.39645	.43793	2.28348	.45889	2.17916	21
40	.37720	2.65109	.39727	2.51715	.41763	2.39449	.43828	2.28167	.45924	2.17749	20
41	.37754	2.64875	.39761	2.51502	.41797	2.39253	.43862	2.27987	.45960	2.17582	19
42	.37787	2.64642	.39795	2.51289	.41831	2.39058	.43897	2.27806	.45995	2.17416	18
43	.37820	2.64410	.39829	2.51076	.41865	2.38863	.43932	2.27626	.46030	2.17249	17
44	.37853	2.64177	.39862	2.50864	.41899	2.38668	.43966	2.27447	.46065	2.17083	16
45	.37887	2.63945	.39896	2.50652	.41933	2.38473	.44001	2.27267	.46101	2.16917	15
46	.37920	2.63714	.39930	2.50440	.41968	2.38279	.44036	2.27088	.46136	2.16751	14
47	.37953	2.63483	.39963	2.50229	.42002	2.38084	.44071	2.26909	.46171	2.16585	13
48	.37986	2.63252	.39997	2.50018	.42036	2.37891	.44105	2.26730	.46206	2.16420	12
49	.38020	2.63021	.40031	2.49807	.42070	2.37697	.44140	2.26552	.46242	2.16255	11
50	.38053	2.62791	.40065	2.49597	.42105	2.37504	.44175	2.26374	.46277	2.16090	10
51	.38086	2.62561	.40098	2.49386	.42139	2.37311	.44210	2.26196	.46312	2.15925	9
52	.38120	2.62332	.40132	2.49177	.42173	2.37118	.44244	2.26018	.46348	2.15760	8
53	.38153	2.62103	.40166	2.48967	.42207	2.36925	.44279	2.25840	.46383	2.15596	7
54	.38186	2.61874	.40200	2.48758	.42242	2.36733	.44314	2.25663	.46418	2.15432	6
55	.38220	2.61646	.40234	2.48549	.42276	2.36541	.44349	2.25486	.46454	2.15268	5
56	.38253	2.61418	.40267	2.48340	.42310	2.36349	.44384	2.25309	.46489	2.15104	4
57	.38286	2.61190	.40301	2.48132	.42345	2.36158	.44418	2.25132	.46525	2.14940	3
58	.38320	2.60963	.40335	2.47924	.42379	2.35967	.44453	2.24956	.46560	2.14777	2
59	.38353	2.60736	.40369	2.47716	.42413	2.35776	.44488	2.24780	.46595	2.14614	1
60	.38386	2.60509	.40403	2.47509	.42447	2.35585	.44523	2.24604	.46631	2.14451	0
′	Cotang	Tang	Cotang	Tang	Cotang	Tang	Cotang	Tang	Cotang	Tang	′
	69°		68°		67°		66°		65°		

′	25° Tang	Cotang	26° Tang	Cotang	27° Tang	Cotang	28° Tang	Cotang	29° Tang	Cotang	′
0	.46631	2.14451	.48773	2.05030	.50953	1.96261	.53171	1.88073	.55431	1.80405	60
1	.46666	2.14288	.48809	2.04879	.50989	1.96120	.53208	1.87941	.55469	1.80281	59
2	.46702	2.14125	.48845	2.04728	.51026	1.95979	.53246	1.87809	.55507	1.80158	58
3	.46737	2.13963	.48881	2.04577	.51063	1.95838	.53283	1.87677	.55545	1.80034	57
4	.46772	2.13801	.48917	2.04426	.51099	1.95698	.53320	1.87546	.55583	1.79911	56
5	.46808	2.13639	.48953	2.04276	.51136	1.95557	.53358	1.87415	.55621	1.79788	55
6	.46843	2.13477	.48989	2.04125	.51173	1.95417	.53395	1.87283	.55659	1.79665	54
7	.46879	2.13316	.49026	2.03975	.51209	1.95277	.53432	1.87152	.55697	1.79542	53
8	.46914	2.13154	.49062	2.03825	.51246	1.95137	.53470	1.87021	.55736	1.79419	52
9	.46950	2.12993	.49098	2.03675	.51283	1.94997	.53507	1.86891	.55774	1.79296	51
10	.46985	2.12832	.49134	2.03526	.51319	1.94858	.53545	1.86760	.55812	1.79174	50
11	.47021	2.12671	.49170	2.03376	.51356	1.94718	.53582	1.86630	.55850	1.79051	49
12	.47056	2.12511	.49206	2.03227	.51393	1.94579	.53620	1.86499	.55888	1.78929	48
13	.47092	2.12350	.49242	2.03078	.51430	1.94440	.53657	1.86369	.55926	1.78807	47
14	.47128	2.12190	.49278	2.02929	.51467	1.94301	.53694	1.86239	.55964	1.78685	46
15	.47163	2.12030	.49315	2.02780	.51503	1.94162	.53732	1.86109	.56003	1.78563	45
16	.47199	2.11871	.49351	2.02631	.51540	1.94023	.53769	1.85979	.56041	1.78441	44
17	.47234	2.11711	.49387	2.02483	.51577	1.93885	.53807	1.85850	.56079	1.78319	43
18	.47270	2.11552	.49423	2.02335	.51614	1.93746	.53844	1.85720	.56117	1.78198	42
19	.47305	2.11392	.49459	2.02187	.51651	1.93608	.53882	1.85591	.56156	1.78077	41
20	.47341	2.11233	.49495	2.02039	.51688	1.93470	.53920	1.85462	.56194	1.77955	40
21	.47377	2.11075	.49532	2.01891	.51724	1.93332	.53957	1.85333	.56232	1.77834	39
22	.47412	2.10916	.49568	2.01743	.51761	1.93195	.53995	1.85204	.56270	1.77713	38
23	.47448	2.10758	.49604	2.01596	.51798	1.93057	.54032	1.85075	.56309	1.77592	37
24	.47483	2.10600	.49640	2.01449	.51835	1.92920	.54070	1.84946	.56347	1.77471	36
25	.47519	2.10442	.49677	2.01302	.51872	1.92782	.54107	1.84818	.56385	1.77351	35
26	.47555	2.10284	.49713	2.01155	.51909	1.92645	.54145	1.84689	.56424	1.77230	34
27	.47590	2.10126	.49749	2.01008	.51946	1.92508	.54183	1.84561	.56462	1.77110	33
28	.47626	2.09969	.49786	2.00862	.51983	1.92371	.54220	1.84433	.56501	1.76990	32
29	.47662	2.09811	.49822	2.00715	.52020	1.92235	.54258	1.84305	.56539	1.76869	31
30	.47698	2.09654	.49858	2.00569	.52057	1.92098	.54296	1.84177	.56577	1.76749	30
31	.47733	2.09498	.49894	2.00423	.52094	1.91962	.54333	1.84049	.56616	1.76629	29
32	.47769	2.09341	.49931	2.00277	.52131	1.91826	.54371	1.83922	.56654	1.76510	28
33	.47805	2.09184	.49967	2.00131	.52168	1.91690	.54409	1.83794	.56693	1.76390	27
34	.47840	2.09028	.50004	1.99986	.52205	1.91554	.54446	1.83667	.56731	1.76271	26
35	.47876	2.08872	.50040	1.99841	.52242	1.91418	.54484	1.83540	.56769	1.76151	25
36	.47912	2.08716	.50076	1.99695	.52279	1.91282	.54522	1.83413	.56808	1.76032	24
37	.47948	2.08560	.50113	1.99550	.52316	1.91147	.54560	1.83286	.56846	1.75913	23
38	.47984	2.08405	.50149	1.99406	.52353	1.91012	.54597	1.83159	.56885	1.75794	22
39	.48019	2.08250	.50185	1.99261	.52390	1.90876	.54635	1.83033	.56923	1.75675	21
40	.48055	2.08094	.50222	1.99116	.52427	1.90741	.54673	1.82906	.56962	1.75556	20
41	.48091	2.07939	.50258	1.98972	.52464	1.90607	.54711	1.82780	.57000	1.75437	19
42	.48127	2.07785	.50295	1.98828	.52501	1.90472	.54748	1.82654	.57039	1.75319	18
43	.48163	2.07630	.50331	1.98684	.52538	1.90337	.54786	1.82528	.57078	1.75200	17
44	.48198	2.07476	.50368	1.98540	.52575	1.90203	.54824	1.82402	.57116	1.75082	16
45	.48234	2.07321	.50404	1.98396	.52613	1.90069	.54862	1.82276	.57155	1.74964	15
46	.48270	2.07167	.50441	1.98253	.52650	1.89935	.54900	1.82150	.57193	1.74846	14
47	.48306	2.07014	.50477	1.98110	.52687	1.89801	.54938	1.82025	.57232	1.74728	13
48	.48342	2.06860	.50514	1.97966	.52724	1.89667	.54975	1.81899	.57271	1.74610	12
49	.48378	2.06706	.50550	1.97823	.52761	1.89533	.55013	1.81774	.57309	1.74492	11
50	.48414	2.06553	.50587	1.97681	.52798	1.89400	.55051	1.81649	.57348	1.74375	10
51	.48450	2.06400	.50623	1.97538	.52836	1.89266	.55089	1.81524	.57386	1.74257	9
52	.48486	2.06247	.50660	1.97395	.52873	1.89133	.55127	1.81399	.57425	1.74140	8
53	.48521	2.06094	.50696	1.97253	.52910	1.89000	.55165	1.81274	.57464	1.74022	7
54	.48557	2.05942	.50733	1.97111	.52947	1.88867	.55203	1.81150	.57503	1.73905	6
55	.48593	2.05790	.50769	1.96969	.52985	1.88734	.55241	1.81025	.57541	1.73788	5
56	.48629	2.05637	.50806	1.96827	.53022	1.88602	.55279	1.80901	.57580	1.73671	4
57	.48665	2.05485	.50843	1.96685	.53059	1.88469	.55317	1.80777	.57619	1.73555	3
58	.48701	2.05333	.50879	1.96544	.53096	1.88337	.55355	1.80653	.57657	1.73438	2
59	.48737	2.05182	.50916	1.96402	.53134	1.88205	.55393	1.80529	.57696	1.73321	1
60	.48773	2.05030	.50953	1.96261	.53171	1.88073	.55431	1.80405	.57735	1.73205	0
′	Cotang	Tang	Cotang	Tang	Cotang	Tang	Cotang	Tang	Cotang	Tang	′
	64°		63°		62°		61°		60°		

′	30° Tang	Cotang	31° Tang	Cotang	32° Tang	Cotang	33° Tang	Cotang	34° Tang	Cotang	′
0	.57735	1.73205	.60086	1.66428	.62487	1.60033	.64941	1.53986	.67451	1.48256	60
1	.57774	1.73089	.60126	1.66318	.62527	1.59930	.64982	1.53888	.67493	1.48163	59
2	.57813	1.72973	.60165	1.66209	.62568	1.59826	.65024	1.53791	.67536	1.48070	58
3	.57851	1.72857	.60205	1.66099	.62608	1.59723	.65065	1.53693	.67578	1.47977	57
4	.57890	1.72741	.60245	1.65990	.62649	1.59620	.65106	1.53595	.67620	1.47885	56
5	.57929	1.72625	.60284	1.65881	.62689	1.59517	.65148	1.53497	.67663	1.47792	55
6	.57968	1.72509	.60324	1.65772	.62730	1.59414	.65189	1.53400	.67705	1.47699	54
7	.58007	1.72393	.60364	1.65663	.62770	1.59311	.65231	1.53302	.67748	1.47607	53
8	.58046	1.72278	.60403	1.65554	.62811	1.59208	.65272	1.53205	.67790	1.47514	52
9	.58085	1.72163	.60443	1.65445	.62852	1.59105	.65314	1.53107	.67832	1.47422	51
10	.58124	1.72047	.60483	1.65337	.62892	1.59002	.65355	1.53010	.67875	1.47330	50
11	.58162	1.71932	.60522	1.65228	.62933	1.58900	.65397	1.52913	.67917	1.47238	49
12	.58201	1.71817	.60562	1.65120	.62973	1.58797	.65438	1.52816	.67960	1.47146	48
13	.58240	1.71702	.60602	1.65011	.63014	1.58695	.65480	1.52719	.68002	1.47053	47
14	.58279	1.71588	.60642	1.64903	.63055	1.58593	.65521	1.52622	.68045	1.46962	46
15	.58318	1.71473	.60681	1.64795	.63095	1.58490	.65563	1.52525	.68088	1.46870	45
16	.58357	1.71358	.60721	1.64687	.63136	1.58388	.65604	1.52429	.68130	1.46778	44
17	.58396	1.71244	.60761	1.64579	.63177	1.58286	.65646	1.52332	.68173	1.46686	43
18	.58435	1.71129	.60801	1.64471	.63217	1.58184	.65688	1.52235	.68215	1.46595	42
19	.58474	1.71015	.60841	1.64363	.63258	1.58083	.65729	1.52139	.68258	1.46503	41
20	.58513	1.70901	.60881	1.64256	.63299	1.57981	.65771	1.52043	.68301	1.46411	40
21	.58552	1.70787	.60921	1.64148	.63340	1.57879	.65813	1.51946	.68343	1.46320	39
22	.58591	1.70673	.60960	1.64041	.63380	1.57778	.65854	1.51850	.68386	1.46229	38
23	.58631	1.70560	.61000	1.63934	.63421	1.57676	.65896	1.51754	.68429	1.46137	37
24	.58670	1.70446	.61040	1.63826	.63462	1.57575	.65938	1.51658	.68471	1.46046	36
25	.58709	1.70332	.61080	1.63719	.63503	1.57474	.65980	1.51562	.68514	1.45955	35
26	.58748	1.70219	.61120	1.63612	.63544	1.57372	.66021	1.51466	.68557	1.45864	34
27	.58787	1.70106	.61160	1.63505	.63584	1.57271	.66063	1.51370	.68600	1.45773	33
28	.58826	1.69992	.61200	1.63398	.63625	1.57170	.66105	1.51275	.68642	1.45682	32
29	.58865	1.69879	.61240	1.63292	.63666	1.57069	.66147	1.51179	.68685	1.45592	31
30	.58905	1.69766	.61280	1.63185	.63707	1.56969	.66189	1.51084	.68728	1.45501	30
31	.58944	1.69653	.61320	1.63079	.63748	1.56868	.66230	1.50988	.68771	1.45410	29
32	.58983	1.69541	.61360	1.62972	.63789	1.56767	.66272	1.50893	.68814	1.45320	28
33	.59022	1.69428	.61400	1.62866	.63830	1.56667	.66314	1.50797	.68857	1.45229	27
34	.59061	1.69316	.61440	1.62760	.63871	1.56566	.66356	1.50702	.68900	1.45139	26
35	.59101	1.69203	.61480	1.62654	.63912	1.56466	.66398	1.50607	.68942	1.45049	25
36	.59140	1.69091	.61520	1.62548	.63953	1.56366	.66440	1.50512	.68985	1.44958	24
37	.59179	1.68979	.61561	1.62442	.63994	1.56265	.66482	1.50417	.69028	1.44868	23
38	.59218	1.68866	.61601	1.62336	.64035	1.56165	.66524	1.50322	.69071	1.44778	22
39	.59258	1.68754	.61641	1.62230	.64076	1.56065	.66566	1.50228	.69114	1.44688	21
40	.59297	1.68643	.61681	1.62125	.64117	1.55966	.66608	1.50133	.69157	1.44598	20
41	.59336	1.68531	.61721	1.62019	.64158	1.55866	.66650	1.50038	.69200	1.44508	19
42	.59376	1.68419	.61761	1.61914	.64199	1.55766	.66692	1.49944	.69243	1.44418	18
43	.59415	1.68308	.61801	1.61808	.64240	1.55666	.66734	1.49849	.69286	1.44329	17
44	.59454	1.68196	.61842	1.61703	.64281	1.55567	.66776	1.49755	.69329	1.44239	16
45	.59494	1.68085	.61882	1.61598	.64322	1.55467	.66818	1.49661	.69372	1.44149	15
46	.59533	1.67974	.61922	1.61493	.64363	1.55368	.66860	1.49566	.69416	1.44060	14
47	.59573	1.67863	.61962	1.61388	.64404	1.55269	.66902	1.49472	.69459	1.43970	13
48	.59612	1.67752	.62003	1.61283	.64446	1.55170	.66944	1.49378	.69502	1.43881	12
49	.59651	1.67641	.62043	1.61179	.64487	1.55071	.66986	1.49284	.69545	1.43792	11
50	.59691	1.67530	.62083	1.61074	.64528	1.54972	.67028	1.49190	.69588	1.43703	10
51	.59730	1.67419	.62124	1.60970	.64569	1.54873	.67071	1.49097	.69631	1.43614	9
52	.59770	1.67309	.62164	1.60865	.64610	1.54774	.67113	1.49003	.69675	1.43525	8
53	.59809	1.67198	.62204	1.60761	.64652	1.54675	.67155	1.48909	.69718	1.43436	7
54	.59849	1.67088	.62245	1.60657	.64693	1.54576	.67197	1.48816	.69761	1.43347	6
55	.59888	1.66978	.62285	1.60553	.64734	1.54478	.67239	1.48722	.69804	1.43258	5
56	.59928	1.66867	.62325	1.60449	.64775	1.54379	.67282	1.48629	.69847	1.43169	4
57	.59967	1.66757	.62366	1.60345	.64817	1.54281	.67324	1.48536	.69891	1.43080	3
58	.60007	1.66647	.62406	1.60241	.64858	1.54183	.67366	1.48442	.69934	1.42992	2
59	.60046	1.66538	.62446	1.60137	.64899	1.54085	.67409	1.48349	.69977	1.42903	1
60	.60086	1.66428	.62487	1.60033	.64941	1.53986	.67451	1.48256	.70021	1.42815	0
′	Cotang	Tang	Cotang	Tang	Cotang	Tang	Cotang	Tang	Cotang	Tang	′
	59°		58°		57°		56°		55°		

′	35° Tang	Cotang	36° Tang	Cotang	37° Tang	Cotang	38° Tang	Cotang	39° Tang	Cotang	′
0	.70021	1.42815	.72654	1.37638	.75355	1.32704	.78129	1.27994	.80978	1.23490	60
1	.70064	1.42726	.72699	1.37554	.75401	1.32624	.78175	1.27917	.81027	1.23416	59
2	.70107	1.42638	.72743	1.37470	.75447	1.32544	.78222	1.27841	.81075	1.23343	58
3	.70151	1.42550	.72788	1.37386	.75492	1.32464	.78269	1.27764	.81123	1.23270	57
4	.70194	1.42462	.72832	1.37302	.75538	1.32384	.78316	1.27688	.81171	1.23196	56
5	.70238	1.42374	.72877	1.37218	.75584	1.32304	.78363	1.27611	.81220	1.23123	55
6	.70281	1.42286	.72921	1.37134	.75629	1.32224	.78410	1.27535	.81268	1.23050	54
7	.70325	1.42198	.72966	1.37050	.75675	1.32144	.78457	1.27458	.81316	1.22977	53
8	.70368	1.42110	.73010	1.36967	.75721	1.32064	.78504	1.27382	.81364	1.22904	52
9	.70412	1.42022	.73055	1.36883	.75767	1.31984	.78551	1.27306	.81413	1.22831	51
10	.70455	1.41934	.73100	1.36800	.75812	1.31904	.78598	1.27230	.81461	1.22758	50
11	.70499	1.41847	.73144	1.36716	.75858	1.31825	.78645	1.27153	.81510	1.22685	49
12	.70542	1.41759	.73189	1.36633	.75904	1.31745	.78692	1.27077	.81558	1.22612	48
13	.70586	1.41672	.73234	1.36549	.75950	1.31666	.78739	1.27001	.81606	1.22539	47
14	.70629	1.41584	.73278	1.36466	.75996	1.31586	.78786	1.26925	.81655	1.22467	46
15	.70673	1.41497	.73323	1.36383	.76042	1.31507	.78834	1.26849	.81703	1.22394	45
16	.70717	1.41409	.73368	1.36300	.76088	1.31427	.78881	1.26774	.81752	1.22321	44
17	.70760	1.41322	.73413	1.36217	.76134	1.31348	.78928	1.26698	.81800	1.22249	43
18	.70804	1.41235	.73457	1.36134	.76180	1.31269	.78975	1.26622	.81849	1.22176	42
19	.70848	1.41148	.73502	1.36051	.76226	1.31190	.79022	1.26546	.81898	1.22104	41
20	.70891	1.41061	.73547	1.35968	.76272	1.31110	.79070	1.26471	.81946	1.22031	40
21	.70935	1.40974	.73592	1.35885	.76318	1.31031	.79117	1.26395	.81995	1.21959	39
22	.70979	1.40887	.73637	1.35802	.76364	1.30952	.79164	1.26319	.82044	1.21886	38
23	.71023	1.40800	.73681	1.35719	.76410	1.30873	.79212	1.26244	.82092	1.21814	37
24	.71066	1.40714	.73726	1.35637	.76456	1.30795	.79259	1.26169	.82141	1.21742	36
25	.71110	1.40627	.73771	1.35554	.76502	1.30716	.79306	1.26093	.82190	1.21670	35
26	.71154	1.40540	.73816	1.35472	.76548	1.30637	.79354	1.26018	.82238	1.21598	34
27	.71198	1.40454	.73861	1.35389	.76594	1.30558	.79401	1.25943	.82287	1.21526	33
28	.71242	1.40367	.73906	1.35307	.76640	1.30480	.79449	1.25867	.82336	1.21454	32
29	.71285	1.40281	.73951	1.35224	.76686	1.30401	.79496	1.25792	.82385	1.21382	31
30	.71329	1.40195	.73996	1.35142	.76733	1.30323	.79544	1.25717	.82434	1.21310	30
31	.71373	1.40109	.74041	1.35060	.76779	1.30244	.79591	1.25642	.82483	1.21238	29
32	.71417	1.40022	.74086	1.34978	.76825	1.30166	.79639	1.25567	.82531	1.21166	28
33	.71461	1.39936	.74131	1.34896	.76871	1.30087	.79686	1.25492	.82580	1.21094	27
34	.71505	1.39850	.74176	1.34814	.76918	1.30009	.79734	1.25417	.82629	1.21023	26
35	.71549	1.39764	.74221	1.34732	.76964	1.29931	.79781	1.25343	.82678	1.20951	25
36	.71593	1.39679	.74267	1.34650	.77010	1.29853	.79829	1.25268	.82727	1.20879	24
37	.71637	1.39593	.74312	1.34568	.77057	1.29775	.79877	1.25193	.82776	1.20808	23
38	.71681	1.39507	.74357	1.34487	.77103	1.29696	.79924	1.25118	.82825	1.20736	22
39	.71725	1.39421	.74402	1.34405	.77149	1.29618	.79972	1.25044	.82874	1.20665	21
40	.71769	1.39336	.74447	1.34323	.77196	1.29541	.80020	1.24969	.82923	1.20593	20
41	.71813	1.39250	.74492	1.34242	.77242	1.29463	.80067	1.24895	.82972	1.20522	19
42	.71857	1.39165	.74538	1.34160	.77289	1.29385	.80115	1.24820	.83022	1.20451	18
43	.71901	1.39079	.74583	1.34079	.77335	1.29307	.80163	1.24746	.83071	1.20379	17
44	.71946	1.38994	.74628	1.33998	.77382	1.29229	.80211	1.24672	.83120	1.20308	16
45	.71990	1.38909	.74674	1.33916	.77428	1.29152	.80258	1.24597	.83169	1.20237	15
46	.72034	1.38824	.74719	1.33835	.77475	1.29074	.80306	1.24523	.83218	1.20166	14
47	.72078	1.38738	.74764	1.33754	.77521	1.28997	.80354	1.24449	.83268	1.20095	13
48	.72122	1.38653	.74810	1.33673	.77568	1.28919	.80402	1.24375	.83317	1.20024	12
49	.72167	1.38568	.74855	1.33592	.77615	1.28842	.80450	1.24301	.83366	1.19953	11
50	.72211	1.38484	.74900	1.33511	.77661	1.28764	.80498	1.24227	.83415	1.19882	10
51	.72255	1.38399	.74946	1.33430	.77708	1.28687	.80546	1.24153	.83465	1.19811	9
52	.72299	1.38314	.74991	1.33349	.77754	1.28610	.80594	1.24079	.83514	1.19740	8
53	.72344	1.38229	.75037	1.33268	.77801	1.28533	.80642	1.24005	.83564	1.19669	7
54	.72388	1.38145	.75082	1.33187	.77848	1.28456	.80690	1.23931	.83613	1.19599	6
55	.72432	1.38060	.75128	1.33107	.77895	1.28379	.80738	1.23858	.83662	1.19528	5
56	.72477	1.37976	.75173	1.33026	.77941	1.28302	.80786	1.23784	.83712	1.19457	4
57	.72521	1.37891	.75219	1.32946	.77988	1.28225	.80834	1.23710	.83761	1.19387	3
58	.72565	1.37807	.75264	1.32865	.78035	1.28148	.80882	1.23637	.83811	1.19316	2
59	.72610	1.37722	.75310	1.32785	.78082	1.28071	.80930	1.23563	.83860	1.19246	1
60	.72654	1.37638	.75355	1.32704	.78129	1.27994	.80978	1.23490	.83910	1.19175	0
′	Cotang	Tang	Cotang	Tang	Cotang	Tang	Cotang	Tang	Cotang	Tang	′
	54°		53°		52°		51°		50°		

′	40° Tang	40° Cotang	41° Tang	41° Cotang	42° Tang	42° Cotang	43° Tang	43° Cotang	44° Tang	44° Cotang	′
0	.83910	1.19175	.86929	1.15037	.90040	1.11061	.93252	1.07237	.96569	1.03553	60
1	.83960	1.19105	.86980	1.14969	.90093	1.10996	.93306	1.07174	.96625	1.03493	59
2	.84009	1.19035	.87031	1.14902	.90146	1.10931	.93360	1.07112	.96681	1.03433	58
3	.84059	1.18964	.87082	1.14834	.90199	1.10867	.93415	1.07049	.96738	1.03372	57
4	.84108	1.18894	.87133	1.14767	.90251	1.10802	.93469	1.06987	.96794	1.03312	56
5	.84158	1.18824	.87184	1.14699	.90304	1.10737	.93524	1.06925	.96850	1.03252	55
6	.84208	1.18754	.87236	1.14632	.90357	1.10672	.93578	1.06862	.96907	1.03192	54
7	.84258	1.18684	.87287	1.14565	.90410	1.10607	.93633	1.06800	.96963	1.03132	53
8	.84307	1.18614	.87338	1.14498	.90463	1.10543	.93688	1.06738	.97020	1.03072	52
9	.84357	1.18544	.87389	1.14430	.90516	1.10478	.93742	1.06676	.97076	1.03012	51
10	.84407	1.18474	.87441	1.14363	.90569	1.10414	.93797	1.06613	.97133	1.02952	50
11	.84457	1.18404	.87492	1.14296	.90621	1.10349	.93852	1.06551	.97189	1.02892	49
12	.84507	1.18334	.87543	1.14229	.90674	1.10285	.93906	1.06489	.97246	1.02832	48
13	.84556	1.18264	.87595	1.14162	.90727	1.10220	.93961	1.06427	.97302	1.02772	47
14	.84606	1.18194	.87646	1.14095	.90781	1.10156	.94016	1.06365	.97359	1.02713	46
15	.84656	1.18125	.87698	1.14028	.90834	1.10091	.94071	1.06303	.97416	1.02653	45
16	.84706	1.18055	.87749	1.13961	.90887	1.10027	.94125	1.06241	.97472	1.02593	44
17	.84756	1.17986	.87801	1.13894	.90940	1.09963	.94180	1.06179	.97529	1.02533	43
18	.84806	1.17916	.87852	1.13828	.90993	1.09899	.94235	1.06117	.97586	1.02474	42
19	.84856	1.17846	.87904	1.13761	.91046	1.09834	.94290	1.06056	.97643	1.02414	41
20	.84906	1.17777	.87955	1.13694	.91099	1.09770	.94345	1.05994	.97700	1.02355	40
21	.84956	1.17708	.88007	1.13627	.91153	1.09706	.94400	1.05932	.97756	1.02295	39
22	.85006	1.17638	.88059	1.13561	.91206	1.09642	.94455	1.05870	.97813	1.02236	38
23	.85057	1.17569	.88110	1.13494	.91259	1.09578	.94510	1.05809	.97870	1.02176	37
24	.85107	1.17500	.88162	1.13428	.91313	1.09514	.94565	1.05747	.97927	1.02117	36
25	.85157	1.17430	.88214	1.13361	.91366	1.09450	.94620	1.05685	.97984	1.02057	35
26	.85207	1.17361	.88265	1.13295	.91419	1.09386	.94676	1.05624	.98041	1.01998	34
27	.85257	1.17292	.88317	1.13228	.91473	1.09322	.94731	1.05562	.98098	1.01939	33
28	.85308	1.17223	.88369	1.13162	.91526	1.09258	.94786	1.05501	.98155	1.01879	32
29	.85358	1.17154	.88421	1.13096	.91580	1.09195	.94841	1.05439	.98213	1.01820	31
30	.85408	1.17085	.88473	1.13029	.91633	1.09131	.94896	1.05378	.98270	1.01761	30
31	.85458	1.17016	.88524	1.12963	.91687	1.09067	.94952	1.05317	.98327	1.01702	29
32	.85509	1.16947	.88576	1.12897	.91740	1.09003	.95007	1.05255	.98384	1.01642	28
33	.85559	1.16878	.88628	1.12831	.91794	1.08940	.95062	1.05194	.98441	1.01583	27
34	.85609	1.16809	.88680	1.12765	.91847	1.08876	.95118	1.05133	.98499	1.01524	26
35	.85660	1.16741	.88732	1.12699	.91901	1.08811	.95173	1.05072	.98556	1.01465	25
36	.85710	1.16672	.88784	1.12633	.91955	1.08749	.95229	1.05010	.98613	1.01406	24
37	.85761	1.16603	.88836	1.12567	.92008	1.08686	.95284	1.04949	.98671	1.01347	23
38	.85811	1.16535	.88888	1.12501	.92062	1.08622	.95340	1.04888	.98728	1.01288	22
39	.85862	1.16466	.88940	1.12435	.92116	1.08559	.95395	1.04827	.98786	1.01229	21
40	.85912	1.16398	.88992	1.12369	.92170	1.08496	.95451	1.04766	.98843	1.01170	20
41	.85963	1.16329	.89045	1.12303	.92224	1.08432	.95506	1.04705	.98901	1.01112	19
42	.86014	1.16261	.89097	1.12238	.92277	1.08369	.95562	1.04644	.98958	1.01053	18
43	.86064	1.16192	.89149	1.12172	.92331	1.08306	.95618	1.04583	.99016	1.00994	17
44	.86115	1.16124	.89201	1.12106	.92385	1.08243	.95673	1.04522	.99073	1.00935	16
45	.86166	1.16056	.89253	1.12041	.92439	1.08179	.95729	1.04461	.99131	1.00876	15
46	.86216	1.15987	.89306	1.11975	.92493	1.08116	.95785	1.04401	.99189	1.00818	14
47	.86267	1.15919	.89358	1.11909	.92547	1.08053	.95841	1.04340	.99247	1.00759	13
48	.86318	1.15851	.89410	1.11844	.92601	1.07990	.95897	1.04279	.99304	1.00701	12
49	.86368	1.15783	.89463	1.11778	.92655	1.07927	.95952	1.04218	.99362	1.00642	11
50	.86419	1.15715	.89515	1.11713	.92709	1.07864	.96008	1.04158	.99420	1.00583	10
51	.86470	1.15647	.89567	1.11648	.92763	1.07801	.96064	1.04097	.99478	1.00525	9
52	.86521	1.15579	.89620	1.11582	.92817	1.07738	.96120	1.04036	.99536	1.00467	8
53	.86572	1.15511	.89672	1.11517	.92872	1.07676	.96176	1.03976	.99594	1.00408	7
54	.86623	1.15443	.89725	1.11452	.92926	1.07613	.96232	1.03915	.99652	1.00350	6
55	.86674	1.15375	.89777	1.11387	.92980	1.07550	.96288	1.03855	.99710	1.00291	5
56	.86725	1.15308	.89830	1.11321	.93034	1.07487	.96344	1.03794	.99768	1.00233	4
57	.86776	1.15240	.89883	1.11256	.93088	1.07425	.96400	1.03734	.99826	1.00175	3
58	.86827	1.15172	.89935	1.11191	.93143	1.07362	.96457	1.03674	.99884	1.00116	2
59	.86878	1.15104	.89988	1.11126	.93197	1.07299	.96513	1.03613	.99942	1.00058	1
60	.86929	1.15037	.90040	1.11061	.93252	1.07237	.96569	1.03553	1.00000	1.00000	0

′	Cotang	Tang	Cotang	Tang	Cotang	Tang	Cotang	Tang	Cotang	Tang	′
	49°		48°		47°		46°		45°		

TRAVERSE TABLES

OR

LATITUDES ᴬᴺᴰ DEPARTURES OF COURSES

CALCULATED TO

THREE DECIMAL PLACES

FOR

EACH QUARTER DEGREE OF BEARING

Bearing.	1		2		3		4		5	Bearing.
	Lat.	Dep.	Lat.	Dep.	Lat.	Dep.	Lat.	Dep.	Lat.	
0°	1.000	0.000	2.000	0.000	3.000	0.000	4.000	0.000	5.000	90°
0¼	1.000	0.004	2.000	0.009	3.000	0.013	4.000	0.017	5.000	89¾
0½	1.000	0.009	2.000	0.017	3.000	0.026	4.000	0.035	5.000	89½
0¾	1.000	0.013	2.000	0.026	3.000	0.039	4.000	0.052	5.000	89¼
1°	1.000	0.017	2.000	0.035	3.000	0.052	3.999	0.070	4.999	89°
1¼	1.000	0.022	2.000	0.044	2.999	0.065	3.999	0.087	4.999	88¾
1½	1.000	0.026	1.999	0.052	2.999	0.079	3.999	0.105	4.998	88½
1¾	1.000	0.031	1.999	0.061	2.999	0.092	3.998	0.122	4.998	88¼
2°	0.999	0.035	1.999	9.070	2.998	0.105	3.998	0.140	4.997	88°
2¼	0.999	0.039	1.998	0.079	2.998	0.118	3.997	0.157	4.996	87¾
2½	0.999	0.044	1.998	0.087	2.997	0.131	3.996	0.174	4.995	87½
2¾	0.999	0.048	1.998	0.096	2.997	0.144	3.995	0.192	4.994	87¼
3°	0.999	0.052	1.997	0.105	2.996	0.157	3.995	0.209	4.993	87°
3¼	0.998	0.057	1.997	0.113	2.995	0.170	3.994	0.227	4.992	86¾
3½	0.998	0.061	1.996	0.122	2.994	0.183	3.993	0.244	4.991	86½
3¾	0.998	0.065	1.996	0.131	2.994	0.196	3.991	0.262	4.989	86¼
4°	0.998	0.070	1.995	0.140	2.993	0.209	3.990	0.279	4.988	86°
4¼	0.997	0.074	1.995	0.148	2.992	0.222	3.989	0.296	4.986	85¾
4½	0.997	0.078	1.994	0.157	2.991	0.235	3.988	0.314	4.985	85½
4¾	0.997	0.083	1.993	0.166	2.990	0.248	3.986	0.331	4.983	85¼
5°	0.996	0.087	1.992	0.174	2.989	0.261	3.985	0.349	4.981	85°
5¼	0.996	0.092	1.992	0.183	2.987	0.275	3.983	0.366	4.979	84¾
5½	0.995	0.096	1.991	0.192	2.986	0.288	3.982	0.383	4.977	84½
5¾	0.995	0.100	1.990	0.200	2.985	0.301	3.980	0.401	4.975	84¼
6°	0.995	0.105	1.989	0.209	2.984	0.314	3.978	0.418	4.973	84°
6¼	0.994	0.109	1.988	0.218	2.982	0.327	3.976	0.435	4.970	83¾
6½	0.994	0.113	1.987	0.226	2.981	0.340	3.974	0.453	4.968	83½
6¾	0.993	0.118	1.986	0.235	2.979	0.353	3.972	0.470	4.965	83¼
7°	0.993	0.122	1.985	0.244	2.978	0.366	3.970	0.487	4.963	83°
7¼	0.992	0.126	1.984	0.252	2.976	0.379	3.968	0.505	4.960	82¾
7½	0.991	0.131	1.983	0.261	2.974	0.392	3.966	0.522	4.957	82½
7¾	0.991	0.135	1.982	0.270	2.973	0.405	3.963	0.539	4.954	82¼
8°	0.990	0.139	1.981	0.278	2.971	0.418	3.961	0.557	4.951	82°
8¼	0.990	0.143	1.979	0.287	2.969	0.430	3.959	0.574	4.948	81¾
8½	0.989	0.148	1.978	0.296	2.967	0.443	3.956	0.591	4.945	81½
8¾	0.988	0.152	1.977	0.304	2.965	0.456	3.953	0.608	4.942	81¼
9°	0.988	0.156	1.975	0.313	2.963	0.469	3.951	0.626	4.938	81°
9¼	0.987	0.161	1.974	0.321	2.961	0.482	3.948	0.643	4.935	80¾
9½	0.986	0.165	1.973	0.330	2.959	0.495	3.945	0.660	4.931	80½
9¾	0.986	0.169	1.971	0.339	2.957	0.508	3.942	0.677	4.928	80¼
10°	0.985	0.174	1.970	0.347	2.954	0.521	3.939	0.695	4.924	80°
10¼	0.984	0.178	1.968	0.356	2.952	0.534	3.936	0.712	4.920	79¾
10½	0.983	0.182	1.967	0.364	2.950	0.547	3.933	0.729	4.916	79½
10¾	0.982	0.187	1.965	0.373	2.947	0.560	3.930	0.746	4.912	79¼
11°	0.982	0.191	1.963	0.382	2.945	0.572	3.927	0.763	4.908	79°
11¼	0.981	0.195	1.962	0.390	2.942	0.585	3.923	0.780	4.904	78¾
11½	0.980	0.199	1.960	0.399	2.940	0.598	3.920	0.797	4.900	78½
11¾	0.979	0.204	1.958	0.407	2.937	0.611	3.916	0.815	4.895	78¼
12°	0.978	0.208	1.956	0.416	2.934	0.624	3.913	0.832	4.891	78°
12¼	0.977	0.212	1.954	0.424	2.932	0.637	3.909	0.849	4.886	77¾
12½	0.976	0.216	1.953	0.433	2.929	0.649	3.905	0.866	4.881	77½
12¾	0.975	0.221	1.951	0.441	2.926	0.662	3.901	0.883	4.877	77¼
13°	0.974	0.225	1.949	0.450	2.923	0.675	3.897	0.900	4.872	77°
	Dep.	Lat.	Dep.	Lat.	Dep.	Lat.	Dep.	Lat.	Dep.	Bearing.
	1		2		3		4		5	

Bearing.	5	6		7		8		9		Bearing.
	Dep.	Lat.	Dep.	Lat.	Dep.	Lat.	Dep.	Lat.	Dep.	
0°	0.000	6.000	0.000	7.000	0.000	8.000	0.000	9.000	0.000	90°
0¼	0.022	6.000	0.026	7.000	0.031	8.000	0.035	9.000	0.039	89¾
0½	0.044	6.000	0.052	7.000	0.061	8.000	0.070	9.000	0.079	89½
0¾	0.065	5.999	0.079	6.999	0.092	7.999	0.105	8.999	0.118	89¼
1°	0.087	5.999	0.105	6.999	0.122	7.999	0.140	8.999	0.157	89°
1¼	0.109	5.999	0.131	6.998	0.153	7.998	0.175	8.998	0.196	88¾
1½	0.131	5.998	0.157	6.998	0.183	7.997	0.209	8.997	0.236	88½
1¾	0.153	5.997	0.183	6.997	0.214	7.996	0.244	8.996	0.275	88¼
2°	0.174	5.996	0.209	6.996	0.244	7.995	0.279	8.995	0.314	88°
2¼	0.196	5.995	0.236	6.995	0.275	7.994	0.314	8.993	0.353	87¾
2½	0.218	5.994	0.262	6.993	0.305	7.992	0.349	8.991	0.393	87½
2¾	0.240	5.993	0.288	6.992	0.336	7.991	0.384	8.990	0.432	87¼
3°	0.262	5.992	0.314	6.990	0.366	7.989	0.419	8.988	0.471	87°
3¼	0.283	5.990	0.340	6.989	0.397	7.987	0.454	8.986	0.510	86¾
3½	0.305	5.989	0.366	6.987	0.427	7.985	0.488	8.983	0.549	86½
3¾	0.327	5.987	0.392	6.985	0.458	7.983	0.523	8.981	0.589	86¼
4°	0.349	5.985	0.419	6.983	0.488	7.981	0.558	8.978	0.628	86°
4¼	0.371	5.984	0.445	6.981	0.519	7.978	0.593	8.975	0.667	85¾
4½	0.392	5.982	0.471	6.978	0.549	7.975	0.628	8.972	0.706	85½
4¾	0.414	5.979	0.497	6.976	0.580	7.973	0.662	8.969	0.745	85¼
5°	0.436	5.977	0.523	6.973	0.610	7.970	0.697	8.966	0.784	85°
5¼	0.458	5.975	0.549	6.971	0.641	7.966	0.732	8.962	0.824	84¾
5½	0.479	5.972	0.575	6.968	0.671	7.963	0.767	8.959	0.863	84½
5¾	0.501	5.970	0.601	6.965	0.701	7.960	0.802	8.955	0.902	84¼
6°	0.523	5.967	0.627	6.962	0.732	7.956	0.836	8.951	0.941	84°
6¼	0.544	5.964	0.653	6.958	0.762	7.952	0.871	8.947	0.980	83¾
6½	0.566	5.961	0.679	6.955	0.792	7.949	0.906	8.942	1.019	83½
6¾	0.588	5.958	0.705	6.951	0.823	7.945	0.940	8.938	1.058	83¼
7°	0.609	5.955	0.731	6.948	0.853	7.940	0.975	8.933	1.097	83°
7¼	0.631	5.952	0.757	6.944	0.883	7.936	1.010	8.928	1.136	82¾
7½	0.653	5.949	0.783	6.940	0.914	7.932	1.044	8.923	1.175	82½
7¾	0.674	5.945	0.809	6.936	0.944	7.927	1.079	8.918	1.214	82¼
8°	0.696	5.942	0.835	6.932	0.974	7.922	1.113	8.912	1.253	82°
8¼	0.717	5.938	0.861	6.928	1.004	7.917	1.148	8.907	1.291	81¾
8½	0.739	5.934	0.887	6.923	1.035	7.912	1.182	8.901	1.330	81½
8¾	0.761	5.930	0.913	6.919	1.065	7.907	1.217	8.895	1.369	81¼
9°	0.782	5.926	0.939	6.914	1.095	7.902	1.251	8.889	1.408	81°
9¼	0.804	5.922	0.964	6.909	1.125	7.896	1.286	8.883	1.447	80¾
9½	0.825	5.918	0.990	6.904	1.155	7.890	1.320	8.877	1.485	80½
9¾	0.847	5.913	1.016	6.899	1.185	7.884	1.355	8.870	1.524	80¼
10°	0.868	5.909	1.042	6.894	1.216	7.878	1.389	8.863	1.563	80°
10¼	0.890	5.904	1.068	6.888	1.246	7.872	1.424	8.856	1.601	79¾
10½	0.911	5.900	1.093	6.883	1.276	7.866	1.458	8.849	1.640	79½
10¾	0.933	5.895	1.119	6.877	1.306	7.860	1.492	8.842	1.679	79¼
11°	0.954	5.890	1.145	6.871	1.336	7.853	1.526	8.835	1.717	79°
11¼	0.975	5.885	1.171	6.866	1.366	7.846	1.561	8.827	1.756	78¾
11½	0.997	5.880	1.196	6.859	1.396	7.839	1.595	8.819	1.794	78½
11¾	1.018	5.874	1.222	6.853	1.425	7.832	1.629	8.811	1.833	78¼
12°	1.040	5.869	1.247	6.847	1.455	7.825	1.663	8.803	1.871	78°
12¼	1.061	5.863	1.273	6.841	1.485	7.818	1.697	8.795	1.910	77¾
12½	1.082	5.858	1.299	6.834	1.515	7.810	1.732	8.787	1.948	77½
12¾	1.103	5.852	1.324	6.827	1.545	7.803	1.766	8.778	1.986	77¼
13°	1.125	5.846	1.350	6.821	1.575	7.795	1.800	8.769	2.025	77°
Bearing.	Lat.	Dep.	Lat.	Dep.	Lat.	Dep.	Lat.	Dep.	Lat.	Bearing.
	5	6		7		8		9		

Bearing.	1		2		3		4		5	Bearing.
	Lat.	Dep.	Lat.	Dep.	Lat.	Dep.	Lat.	Dep.	Lat.	
13°	0.974	0.225	1.949	0.450	2.923	0.675	3.897	0.900	4.872	77°
13¼	0.973	0.229	1.947	0.458	2.920	0.688	3.894	0.917	4.867	76¾
13½	0.972	0.233	1.945	0.467	2.917	0.700	3.889	0.934	4.862	76½
13¾	0.971	0.238	1.943	0.475	2.914	0.713	3.885	0.951	4.857	76¼
14°	0.970	0.242	1.941	0.484	2.911	0.726	3.881	0.968	4.851	76°
14¼	0.969	0.246	1.938	0.492	2.908	0.738	3.877	0.985	4.846	75¾
14½	0.968	0.250	1.936	0.501	2.904	0.751	3.873	1.002	4.841	75½
14¾	0.967	0.255	1.934	0.509	2.901	0.764	3.868	1.018	4.835	75¼
15°	0.966	0.259	1.932	0.518	2.898	0.776	3.864	1.035	4.830	75°
15¼	0.965	0.263	1.930	0.526	2.894	0.789	3.859	1.052	4.824	74¾
15½	0.964	0.267	1.927	0.534	2.891	0.802	3.855	1.069	4.818	74½
15¾	0.962	0.271	1.925	0.543	2.887	0.814	3.850	1.086	4.812	74¼
16°	0.961	0.276	1.923	0.551	2.884	0.827	3.845	1.103	4.806	74°
16¼	0.960	0.280	1.920	0.560	2.880	0.839	3.840	1.119	4.800	73¾
16½	0.959	0.284	1.918	0.568	2.876	0.852	3.835	1.136	4.794	73½
16¾	0.958	0.288	1.915	0.576	2.873	0.865	3.830	1.153	4.788	73¼
17°	0.956	0.292	1.913	0.585	2.869	0.877	3.825	1.169	4.782	73°
17¼	0.955	0.297	1.910	0.593	2.865	0.890	3.820	1.186	4.775	72¾
17½	0.954	0.301	1.907	0.601	2.861	0.902	3.815	1.203	4.769	72½
17¾	0.952	0.305	1.905	0.610	2.857	0.915	3.810	1.220	4.762	72¼
18°	0.951	0.309	1.902	0.618	2.853	0.927	3.804	1.236	4.755	72°
18¼	0.950	0.313	1.899	0.626	2.849	0.939	3.799	1.253	4.748	71¾
18½	0.948	0.317	1.897	0.635	2.845	0.952	3.793	1.269	4.742	71½
18¾	0.947	0.321	1.894	0.643	2.841	0.964	3.788	1.286	4.735	71¼
19°	0.946	0.326	1.891	0.651	2.837	0.977	3.782	1.302	4.728	71°
19¼	0.944	0.330	1.888	0.659	2.832	0.989	3.776	1.319	4.720	70¾
19½	0.943	0.334	1.885	0.668	2.828	1.001	3.771	1.335	4.713	70½
19¾	0.941	0.338	1.882	0.676	2.824	1.014	3.765	1.352	4.706	70¼
20°	0.940	0.342	1.879	0.684	2.819	1.026	3.759	1.368	4.698	70°
20¼	0.938	0.346	1.876	0.692	2.815	1.038	3.753	1.384	4.691	69¾
20½	0.937	0.350	1.873	0.700	2.810	1.051	3.747	1.401	4.683	69½
20¾	0.935	0.354	1.870	0.709	2.805	1.063	3.741	1.417	4.676	69¼
21°	0.934	0.358	1.867	0.717	2.801	1.075	3.734	1.433	4.668	69°
21¼	0.932	0.362	1.864	0.725	2.796	1.087	3.728	1.450	4.660	68¾
21½	0.930	0.367	1.861	0.733	2.791	1.100	3.722	1.466	4.652	68½
21¾	0.929	0.371	1.858	0.741	2.786	1.112	3.715	1.482	4.644	68¼
22°	0.927	0.375	1.854	0.749	2.782	1.124	3.709	1.498	4.636	68°
22¼	0.926	0.379	1.851	0.757	2.777	1.136	3.702	1.515	4.628	67¾
22½	0.924	0.383	1.848	0.765	2.772	1.148	3.696	1.531	4.619	67½
22¾	0.922	0.387	1.844	0.773	2.767	1.160	3.689	1.547	4.611	67¼
23°	0.921	0.391	1.841	0.781	2.762	1.172	3.682	1.563	4.603	67°
23¼	0.919	0.395	1.838	0.789	2.756	1.184	3.675	1.579	4.594	66¾
23½	0.917	0.399	1.834	0.797	2.751	1.196	3.668	1.595	4.585	66½
23¾	0.915	0.403	1.831	0.805	2.746	1.208	3.661	1.611	4.577	66¼
24°	0.914	0.407	1.827	0.813	2.741	1.220	3.654	1.627	4.568	66°
24¼	0.912	0.411	1.824	0.821	2.735	1.232	3.647	1.643	4.559	65¾
24½	0.910	0.415	1.820	0.829	2.730	1.244	3.640	1.659	4.550	65½
24¾	0.908	0.419	1.816	0.837	2.724	1.256	3.633	1.675	4.541	65¼
25°	0.906	0.423	1.813	0.845	2.719	1.268	3.625	1.690	4.532	65°
25¼	0.904	0.427	1.809	0.853	2.713	1.280	3.618	1.706	4.522	64¾
25½	0.903	0.431	1.805	0.861	2.708	1.292	3.610	1.722	4.513	64½
25¾	0.901	0.434	1.801	0.869	2.702	1.303	3.603	1.738	4.503	64¼
26°	0.899	0.438	1.798	0.877	2.696	1.315	3.595	1.753	4.494	64°
	Dep.	Lat.	Dep.	Lat.	Dep.	Lat.	Dep.	Lat.	Dep.	Bearing.
	1		2		3		4		5	

Bearing	5	6		7		8		9		Bearing
	Dep.	Lat.	Dep.	Lat.	Dep.	Lat.	Dep.	Lat.	Dep.	
13°	1.125	5.846	1.350	6.821	1.575	7.795	1.800	8.769	2.025	77°
13¼	1.146	5.840	1.375	6.814	1.604	7.787	1.834	8.760	2.063	76¾
13½	1.167	5.834	1.401	6.807	1.634	7.779	1.868	8.751	2.101	76½
13¾	1.188	5.828	1.426	6.799	1.664	7.771	1.902	8.742	2.139	76¼
14°	1.210	5.822	1.452	6.792	1.693	7.762	1.935	8.733	2.177	76°
14¼	1.231	5.815	1.477	6.785	1.723	7.754	1.969	8.723	2.215	75¾
14½	1.252	5.809	1.502	6.777	1.753	7.745	2.003	8.713	2.253	75½
14¾	1.273	5.802	1.528	6.769	1.782	7.736	2.037	8.703	2.291	75¼
15°	1.294	5.796	1.553	6.761	1.812	7.727	2.071	8.693	2.329	75°
15¼	1.315	5.789	1.578	6.754	1.841	7.718	2.104	8.683	2.367	74¾
15½	1.336	5.782	1.603	6.745	1.871	7.709	2.138	8.673	2.405	74½
15¾	1.357	5.775	1.629	6.737	1.900	7.700	2.172	8.662	2.443	74¼
16°	1.378	5.768	1.654	6.729	1.929	7.690	2.205	8.651	2.481	74°
16¼	1.399	5.760	1.679	6.720	1.959	7.680	2.239	8.640	2.518	73¾
16½	1.420	5.753	1.704	6.712	1.988	7.671	2.272	8.629	2.556	73½
16¾	1.441	5.745	1.729	6.703	2.017	7.661	2.306	8.618	2.594	73¼
17°	1.462	5.738	1.754	6.694	2.047	7.650	2.339	8.607	2.631	73°
17¼	1.483	5.730	1.779	6.685	2.076	7.640	2.372	8.595	2.669	72¾
17½	1.504	5.722	1.804	6.676	2.105	7.630	2.406	8.583	2.706	72½
17¾	1.524	5.714	1.829	6.667	2.134	7.619	2.439	8.572	2.744	72¼
18°	1.545	5.706	1.854	6.657	2.163	7.608	2.472	8.560	2.781	72°
18¼	1.566	5.698	1.879	6.648	2.192	7.598	2.505	8.547	2.818	71¾
18½	1.587	5.690	1.904	6.638	2.221	7.587	2.538	8.535	2.856	71½
18¾	1.607	5.682	1.929	6.629	2.250	7.575	2.572	8.522	2.893	71¼
19°	1.628	5.673	1.953	6.619	2.279	7.564	2.605	8.510	2.930	71°
19¼	1.648	5.665	1.978	6.609	2.308	7.553	2.638	8.497	2.967	70¾
19½	1.669	5.656	2.003	6.598	2.337	7.541	2.670	8.484	3.004	70½
19¾	1.690	5.647	2.028	6.588	2.365	7.529	2.703	8.471	3.041	70¼
20°	1.710	5.638	2.052	6.578	2.394	7.518	2.736	8.457	3.078	70°
20¼	1.731	5.629	2.077	6.567	2.423	7.506	2.769	8.444	3.115	69¾
20½	1.751	5.620	2.101	6.557	2.451	7.493	2.802	8.430	3.152	69½
20¾	1.771	5.611	2.126	6.546	2.480	7.481	2.834	8.416	3.189	69¼
21°	1.792	5.601	2.150	6.535	2.509	7.469	2.867	8.402	3.225	69°
21¼	1.812	5.592	2.175	6.524	2.537	7.456	2.900	8.388	3.262	68¾
21½	1.833	5.582	2.199	6.513	2.566	7.443	2.932	8.374	3.299	68½
21¾	1.853	5.573	2.223	6.502	2.594	7.430	2.964	8.359	3.335	68¼
22°	1.873	5.563	2.248	6.490	2.622	7.417	2.997	8.345	3.371	68°
22¼	1.893	5.553	2.272	6.479	2.651	7.404	3.029	8.330	3.408	67¾
22½	1.913	5.543	2.296	6.467	2.679	7.391	3.061	8.315	3.444	67½
22¾	1.934	5.533	2.320	6.455	2.707	7.378	3.094	8.300	3.480	67¼
23°	1.954	5.523	2.344	6.444	2.735	7.364	3.126	8.285	3.517	67°
23¼	1.974	5.513	2.368	6.432	2.763	7.350	3.158	8.269	3.553	66¾
23½	1.994	5.502	2.392	6.419	2.791	7.336	3.190	8.254	3.589	66½
23¾	2.014	5.492	2.416	6.407	2.819	7.322	3.222	8.238	3.625	66¼
24°	2.034	5.481	2.440	6.395	2.847	7.308	3.254	8.222	3.661	66°
24¼	2.054	5.471	2.464	6.382	2.875	7.294	3.286	8.206	3.696	65¾
24½	2.073	5.460	2.488	6.370	2.903	7.280	3.318	8.190	3.732	65½
24¾	2.093	5.449	2.512	6.357	2.931	7.265	3.349	8.173	3.768	65¼
25°	2.113	5.438	2.536	6.344	2.958	7.250	3.381	8.157	3.804	65°
25¼	2.133	5.427	2.559	6.331	2.986	7.236	3.413	8.140	3.839	64¾
25½	2.153	5.416	2.583	6.318	3.014	7.221	3.444	8.123	3.875	64½
25¾	2.172	5.404	2.607	6.305	3.041	7.206	3.476	8.106	3.910	64¼
26°	2.192	5.393	2.630	6.292	3.069	7.190	3.507	8.089	3.945	64°
Bearing	Lat.	Dep.	Lat.	Dep.	Lat.	Dep.	Lat.	Dep.	Lat.	Bearing
	5	6		7		8		9		

Bearing	1		2		3		4		5	Bearing
	Lat.	Dep.	Lat.	Dep.	Lat.	Dep.	Lat.	Dep.	Lat.	
26°	0.899	0.438	1.798	0.877	2.696	1.315	3.595	1.753	4.494	64°
26¼	0.897	0.442	1.794	0.885	2.691	1.327	3.587	1.769	4.484	63¾
26½	0.895	0.446	1.790	0.892	2.685	1.339	3.580	1.785	4.475	63½
26¾	0.893	0.450	1.786	0.900	2.679	1.350	3.572	1.800	4.465	63¼
27	0.891	0.454	1.782	0.908	2.673	1.362	3.564	1.816	4.455	63°
27¼	0.889	0.458	1.778	0.916	2.667	1.374	3.556	1.831	4.445	62¾
27½	0.887	0.462	1.774	0.923	2.661	1.385	3.548	1.847	4.435	62½
27¾	0.885	0.466	1.770	0.931	2.655	1.397	3.540	1.862	4.425	62¼
28	0.883	0.469	1.766	0.939	2.649	1.408	3.532	1.878	4.415	62°
28¼	0.881	0.473	1.762	0.947	2.643	1.420	3.524	1.893	4.404	61¾
28½	0.879	0.477	1.758	0.954	2.636	1.431	3.515	1.909	4.394	61½
28¾	0.877	0.481	1.753	0.962	2.630	1.443	3.507	1.924	4.384	61¼
29	0.875	0.485	1.749	0.970	2.624	1.454	3.498	1.939	4.373	61°
29¼	0.872	0.489	1.745	0.977	2.617	1.466	3.490	1.954	4.362	60¾
29½	0.870	0.492	1.741	0.985	2.611	1.477	3.481	1.970	4.352	60½
29¾	0.868	0.496	1.736	0.992	2.605	1.489	3.473	1.985	4.341	60¼
30°	0.866	0.500	1.732	1.000	2.598	1.500	3.464	2.000	4.330	60°
30¼	0.864	0.504	1.728	1.008	2.592	1.511	3.455	2.015	4.319	59¾
30½	0.862	0.508	1.723	1.015	2.585	1.523	3.447	2.030	4.308	59½
30¾	0.859	0.511	1.719	1.023	2.578	1.534	3.438	2.045	4.297	59¼
31	0.857	0.515	1.714	1.030	2.572	1.545	3.429	2.060	4.286	59°
31¼	0.855	0.519	1.710	1.038	2.565	1.556	3.420	2.075	4.275	58¾
31½	0.853	0.522	1.705	1.045	2.558	1.567	3.411	2.090	4.263	58½
31¾	0.850	0.526	1.701	1.052	2.551	1.579	3.401	2.105	4.252	58¼
32	0.848	0.530	1.696	1.060	2.544	1.590	3.392	2.120	4.240	58°
32¼	0.846	0.534	1.691	1.067	2.537	1.601	3.383	2.134	4.229	57¾
32½	0.843	0.537	1.687	1.075	2.530	1.612	3.374	2.149	4.217	57½
32¾	0.841	0.541	1.682	1.082	2.523	1.623	3.364	2.164	4.205	57¼
33	0.839	0.545	1.677	1.089	2.516	1.634	3.355	2.179	4.193	57°
33¼	0.836	0.548	1.673	1.097	2.509	1.645	3.345	2.193	4.181	56¾
33½	0.834	0.552	1.668	1.104	2.502	1.656	3.336	2.208	4.169	56½
33¾	0.831	0.556	1.663	1.111	2.494	1.667	3.326	2.222	4.157	56¼
34	0.829	0.559	1.658	1.118	2.487	1.678	3.316	2.237	4.145	56°
34¼	0.827	0.563	1.653	1.126	2.480	1.688	3.306	2.251	4.133	55¾
34½	0.824	0.566	1.648	1.133	2.472	1.699	3.297	2.266	4.121	55½
34¾	0.822	0.570	1.643	1.140	2.465	1.710	3.287	2.280	4.108	55¼
35	0.819	0.574	1.638	1.147	2.457	1.721	3.277	2.294	4.096	55°
35¼	0.817	0.577	1.633	1.154	2.450	1.731	3.267	2.309	4.083	54¾
35½	0.814	0.581	1.628	1.161	2.442	1.742	3.257	2.323	4.071	54½
35¾	0.812	0.584	1.623	1.168	2.435	1.753	3.246	2.337	4.058	54¼
36	0.809	0.588	1.618	1.176	2.427	1.763	3.236	2.351	4.045	54°
36¼	0.806	0.591	1.613	1.183	2.419	1.774	3.226	2.365	4.032	53¾
36½	0.804	0.595	1.608	1.190	2.412	1.784	3.215	2.379	4.019	53½
36¾	0.801	0.598	1.603	1.197	2.404	1.795	3.205	2.393	4.006	53¼
37	0.799	0.602	1.597	1.204	2.396	1.805	3.195	2.407	3.993	53°
37¼	0.796	0.605	1.592	1.211	2.388	1.816	3.184	2.421	3.980	52¾
37½	0.793	0.609	1.587	1.218	2.380	1.826	3.173	2.435	3.967	52½
37¾	0.791	0.612	1.581	1.224	2.372	1.837	3.163	2.449	3.953	52¼
38	0.788	0.616	1.576	1.231	2.364	1.847	3.152	2.463	3.940	52°
38¼	0.785	0.619	1.571	1.238	2.356	1.857	3.141	2.476	3.927	51¾
38½	0.783	0.623	1.565	1.245	2.348	1.868	3.130	2.490	3.913	51½
38¾	0.780	0.626	1.560	1.252	2.340	1.878	3.120	2.504	3.899	51¼
39	0.777	0.629	1.554	1.259	2.331	1.888	3.109	2.517	3.886	51°
	Dep.	Lat.	Dep.	Lat.	Dep.	Lat.	Dep.	Lat.	Dep.	Bearing
	1		2		3		4		5	

Bearing	5 Dep.	6 Lat.	6 Dep.	7 Lat.	7 Dep.	8 Lat.	8 Dep.	9 Lat.	9 Dep.	Bearing
26°	2.192	5.393	2.630	6.292	3.069	7.190	3.507	8.089	3.945	64°
26¼	2.211	5.381	2.654	6.278	3.096	7.175	3.538	8.072	3.981	63¾
26½	2.231	5.370	2.677	6.265	3.123	7.160	3.570	8.054	4.016	63½
26¾	2.250	5.358	2.701	6.251	3.151	7.144	3.601	8.037	4.051	63¼
27°	2.270	5.346	2.724	6.237	3.178	7.128	3.632	8.019	4.086	63°
27¼	2.289	5.334	2.747	6.223	3.205	7.112	3.663	8.001	4.121	62¾
27½	2.309	5.322	2.770	6.209	3.232	7.096	3.694	7.983	4.156	62½
27¾	2.328	5.310	2.794	6.195	3.259	7.080	3.725	7.965	4.190	62¼
28°	2.347	5.298	2.817	6.181	3.286	7.064	3.756	7.947	4.225	62°
28¼	2.367	5.285	2.840	6.166	3.313	7.047	3.787	7.928	4.260	61¾
28½	2.386	5.273	2.863	6.152	3.340	7.031	3.817	7.909	4.294	61½
28¾	2.405	5.260	2.886	6.137	3.367	7.014	3.848	7.891	4.329	61¼
29°	2.424	5.248	2.909	6.122	3.394	6.997	3.878	7.872	4.363	61°
29¼	2.443	5.235	2.932	6.107	3.420	6.980	3.909	7.852	4.398	60¾
29½	2.462	5.222	2.955	6.093	3.447	6.963	3.939	7.833	4.432	60½
29¾	2.481	5.209	2.977	6.077	3.474	6.946	3.970	7.814	4.466	60¼
30°	2.500	5.196	3.000	6.062	3.500	6.928	4.000	7.794	4.500	60
30¼	2.519	5.183	3.023	6.047	3.526	6.911	4.030	7.775	4.534	59¾
30½	2.538	5.170	3.045	6.031	3.553	6.893	4.060	7.755	4.568	59½
30¾	2.556	5.156	3.068	6.016	3.579	6.875	4.090	7.735	4.602	59¼
31°	2.575	5.143	3.090	6.000	3.605	6.857	4.120	7.715	4.635	59°
31¼	2.594	5.129	3.113	5.984	3.631	6.839	4.150	7.694	4.669	58¾
31½	2.612	5.116	3.135	5.968	3.657	6.821	4.180	7.674	4.702	58½
31¾	2.631	5.102	3.157	5.952	3.683	6.803	4.210	7.653	4.736	58¼
32°	2.650	5.088	3.180	5.936	3.709	6.784	4.239	7.632	4.769	58°
32¼	2.668	5.074	3.202	5.920	3.735	6.766	4.269	7.612	4.802	57¾
32½	2.686	5.060	3.224	5.904	3.761	6.747	4.298	7.591	4.836	57½
32¾	2.705	5.046	3.246	5.887	3.787	6.728	4.328	7.569	4.869	57¼
33°	2.723	5.032	3.268	5.871	3.812	6.709	4.357	7.548	4.902	57
33¼	2.741	5.018	3.290	5.854	3.838	6.690	4.386	7.527	4.935	56¾
33½	2.760	5.003	3.312	5.837	3.864	6.671	4.416	7.505	4.967	56½
33¾	2.778	4.989	3.333	5.820	3.889	6.652	4.445	7.483	5.000	56¼
34°	2.796	4.974	3.355	5.803	3.914	6.632	4.474	7.461	5.033	56°
34¼	2.814	4.960	3.377	5.786	3.940	6.613	4.502	7.439	5.065	55¾
34½	2.832	4.945	3.398	5.769	3.965	6.593	4.531	7.417	5.098	55½
34¾	2.850	4.930	3.420	5.752	3.990	6.573	4.560	7.395	5.130	55¼
35°	2.868	4.915	3.441	5.734	4.015	6.553	4.589	7.372	5.162	55
35¼	2.886	4.900	3.463	5.716	4.040	6.533	4.617	7.350	5.194	54¾
35½	2.904	4.885	3.484	5.699	4.065	6.513	4.646	7.327	5.226	54½
35¾	2.921	4.869	3.505	5.681	4.090	6.493	4.674	7.304	5.258	54¼
36°	2.939	4.854	3.527	5.663	4.115	6.472	4.702	7.281	5.290	54
36¼	2.957	4.839	3.548	5.645	4.139	6.452	4.730	7.258	5.322	53¾
36½	2.974	4.823	3.569	5.627	4.164	6.431	4.759	7.235	5.353	53½
36¾	2.992	4.808	3.590	5.609	4.188	6.410	4.787	7.211	5.385	53¼
37°	3.009	4.792	3.611	5.590	4.213	6.389	4.815	7.188	5.416	53°
37¼	3.026	4.776	3.632	5.572	4.237	6.368	4.842	7.164	5.448	52¾
37½	3.044	4.760	3.653	5.554	4.261	6.347	4.870	7.140	5.479	52½
37¾	3.061	4.744	3.673	5.535	4.286	6.326	4.898	7.116	5.510	52¼
38°	3.078	4.728	3.694	5.516	4.310	6.304	4.925	7.092	5.541	52°
38¼	3.095	4.712	3.715	5.497	4.334	6.283	4.953	7.068	5.572	51¾
38½	3.113	4.696	3.735	5.478	4.358	6.261	4.980	7.043	5.603	51½
38¾	3.130	4.679	3.756	5.459	4.381	6.239	5.007	7.019	5.633	51¼
39°	3.147	4.663	3.776	5.440	4.405	6.217	5.035	6.994	5.664	51°
Bearing	Lat.	Dep.	Lat.	Dep.	Lat.	Dep.	Lat.	Dep.	Lat.	Bearing
	5	6		7		8		9		

Bearing	1 Lat.	1 Dep.	2 Lat.	2 Dep.	3 Lat.	3 Dep.	4 Lat.	4 Dep.	5 Lat.	Bearing
39°	0.777	0.629	1.554	1.259	2.331	1.888	3.109	2.517	3.886	51°
39¼	0.774	0.633	1.549	1.265	2.323	1.898	3.098	2.531	3.872	50¾
39½	0.772	0.636	1.543	1.272	2.315	1.908	3.086	2.544	3.858	50½
39¾	0.769	0.639	1.538	1.279	2.307	1.918	3.075	2.558	3.844	50¼
40°	0.766	0.643	1.532	1.286	2.298	1.928	3.064	2.571	3.830	50°
40¼	0.763	0.646	1.526	1.292	2.290	1.938	3.053	2.584	3.816	49¾
40½	0.760	0.649	1.521	1.299	2.281	1.948	3.042	2.598	3.802	49½
40¾	0.758	0.653	1.515	1.306	2.273	1.958	3.030	2.611	3.788	49¼
41°	0.755	0.656	1.509	1.312	2.264	1.968	3.019	2.624	3.774	49°
41¼	0.752	0.659	1.504	1.319	2.256	1.978	3.007	2.637	3.759	48¾
41½	0.749	0.663	1.498	1.325	2.247	1.988	2.996	2.650	3.745	48½
41¾	0.746	0.666	1.492	1.332	2.238	1.998	2.984	2.664	3.730	48¼
42°	0.743	0.669	1.486	1.338	2.229	2.007	2.973	2.677	3.716	48°
42¼	0.740	0.672	1.480	1.345	2.221	2.017	2.961	2.689	3.701	47¾
42½	0.737	0.676	1.475	1.351	2.212	2.027	2.949	2.702	3.686	47½
42¾	0.734	0.679	1.469	1.358	2.203	2.036	2.937	2.715	3.672	47¼
43°	0.731	0.682	1.463	1.364	2.194	2.046	2.925	2.728	3.657	47°
43¼	0.728	0.685	1.457	1.370	2.185	2.056	2.913	2.741	3.642	46¾
43½	0.725	0.688	1.451	1.377	2.176	2.065	2.901	2.753	3.627	46½
43¾	0.722	0.692	1.445	1.383	2.167	2.075	2.889	2.766	3.612	46¼
44°	0.719	0.695	1.439	1.389	2.158	2.084	2.877	2.779	3.597	46°
44¼	0.716	0.698	1.433	1.396	2.149	2.093	2.865	2.791	3.582	45¾
44½	0.713	0.701	1.427	1.402	2.140	2.103	2.853	2.804	3.566	45½
44¾	0.710	0.704	1.420	1.408	2.131	2.112	2.841	2.816	3.551	45¼
45°	0.707	0.707	1.414	1.414	2.121	2.121	2.828	2.828	3.536	45°
B'ring	Dep.	Lat.	Dep.	Lat.	Dep.	Lat.	Dep.	Lat.	Dep.	B'ring

Bearing	5 Dep.	6 Lat.	6 Dep.	7 Lat.	7 Dep.	8 Lat.	8 Dep.	9 Lat.	9 Dep.	Bearing
39°	3.147	4.663	3.776	5.440	4.405	6.217	5.035	6.994	5.664	51°
39¼	3.164	4.646	3.796	5.421	4.429	6.195	5.062	6.970	5.694	50¾
39½	3.180	4.630	3.816	5.401	4.453	6.173	5.089	6.945	5.725	50½
39¾	3.197	4.613	3.837	5.382	4.476	6.151	5.116	6.920	5.755	50¼
40°	3.214	4.596	3.857	5.362	4.500	6.128	5.142	6.894	5.785	50°
40¼	3.231	4.579	3.877	5.343	4.523	6.106	5.169	6.869	5.815	49¾
40½	3.247	4.562	3.897	5.323	4.546	6.083	5.196	6.844	5.845	49½
40¾	3.264	4.545	3.917	5.303	4.569	6.061	5.222	6.818	5.875	49¼
41°	3.280	4.528	3.936	5.283	4.592	6.038	5.248	6.792	5.905	49°
41¼	3.297	4.511	3.956	5.263	4.615	6.015	5.275	6.767	5.934	48¾
41½	3.313	4.494	3.976	5.243	4.638	5.992	5.301	6.741	5.964	48½
41¾	3.329	4.476	3.995	5.222	4.661	5.968	5.327	6.715	5.993	48¼
42°	3.346	4.459	4.015	5.202	4.684	5.945	5.353	6.688	6.022	48°
42¼	3.362	4.441	4.034	5.182	4.707	5.922	5.379	6.662	6.051	47¾
42½	3.378	4.424	4.054	5.161	4.729	5.898	5.405	6.635	6.080	47½
42¾	3.394	4.406	4.073	5.140	4.752	5.875	5.430	6.609	6.109	47¼
43°	3.410	4.388	4.092	5.119	4.774	5.851	5.456	6.582	6.138	47°
43¼	3.426	4.370	4.111	5.099	4.796	5.827	5.481	6.555	6.167	46¾
43½	3.442	4.352	4.130	5.078	4.818	5.803	5.507	6.528	6.195	46½
43¾	3.458	4.334	4.149	5.057	4.841	5.779	5.532	6.501	6.224	46¼
44°	3.473	4.316	4.168	5.035	4.863	5.755	5.557	6.474	6.252	46°
44¼	3.489	4.298	4.187	5.014	4.885	5.730	5.582	6.447	6.280	45¾
44½	3.505	4.280	4.206	4.993	4.906	5.706	5.607	6.419	6.308	45½
44¾	3.520	4.261	4.224	4.971	4.928	5.681	5.632	6.392	6.336	45¼
45°	3.536	4.243	4.243	4.950	4.950	5.657	5.657	6.364	6.364	45°
B'ring	Lat.	Dep.	Lat.	Dep.	Lat.	Dep.	Lat.	Dep.	Lat.	B'ring

TABLE OF

HORIZONTAL DISTANCES AND DIFFERENCES OF ELEVATION FOR STADIA MEASUREMENTS.

The formulas used in the computation of the following tables furnish expressions for *horizontal distances* and *differences of elevation* for stadia measurements with the conditions that the stadia rod *be held vertical* and the stadia wires be *equidistant* from the center wire. The formulas used are as follows: For the horizontal distance

$$D = c \cos n + a k \cos^2 n, \qquad \textbf{(94.)} \quad \text{Art. } \textbf{1301.}$$

in which D = the corrected distance ; c = the constant ; $a k$ = the stadia distance, and n = the vertical angle.

For the difference of elevation, the following formula is used:

$$E = c \sin n + a k \frac{\sin 2n}{2}. \qquad \textbf{(95.)} \quad \text{Art. } \textbf{1301.}$$

For application of tables see Art. **1301.**

HORIZONTAL DISTANCES AND DIFFERENCES OF ELEVATION FOR STADIA MEASUREMENTS.

Minutes.	0°		1°		2°		3°	
	Hor. Dist.	Diff. Elev.	Hor. Dist.	Diff. Elev.	Hor. Dist.	Diff. Elev.	Hor. Dist.	Diff. Elev.
0′	100.00	.00	99.97	1.74	99.88	3.49	99.73	5.23
2	100.00	.06	99.97	1.80	99.87	3.55	99.72	5.28
4	100.00	.12	99.97	1.86	99.87	3.60	99.71	5.34
6	100.00	.17	99.96	1.92	99.87	3.66	99.71	5.40
8	100.00	.23	99.96	1.98	99.86	3.72	99.70	5.46
10	100.00	.29	99.96	2.04	99.86	3.78	99.69	5.52
12	100.00	.35	99.96	2.09	99.85	3.84	99.69	5.57
14	100.00	.41	99.95	2.15	99.85	3.90	99.68	5.63
16	100.00	.47	99.95	2.21	99.84	3.95	99.68	5.69
18	100.00	.52	99.95	2.27	99.84	4.01	99.67	5.75
20	100.00	.58	99.95	2.33	99.83	4.07	99.66	5.80
22	100.00	.64	99.94	2.38	99.83	4.13	99.66	5.86
24	100.00	.70	99.94	2.44	99.82	4.18	99.65	5.92
26	99.99	.76	99.94	2.50	99.82	4.24	99.64	5.98
28	99.99	.81	99.93	2.56	99.81	4.30	99.63	6.04
30	99.99	.87	99.93	2.62	99.81	4.36	99.63	6.09
32	99.99	.93	99.93	2.67	99.80	4.42	99.62	6.15
34	99.99	.99	99.93	2.73	99.80	4.48	99.62	6.21
36	99.99	1.05	99.92	2.79	99.79	4.53	99.61	6.27
38	99.99	1.11	99.92	2.85	99.79	4.59	99.60	6.33
40	99.99	1.16	99.92	2.91	99.78	4.65	99.59	6.38
42	99.99	1.22	99.91	2.97	99.78	4.71	99.59	6.44
44	99.98	1.28	99.91	3.02	99.77	4.76	99.58	6.50
46	99.98	1.34	99.90	3.08	99.77	4.82	99.57	6.56
48	99.98	1.40	99.90	3.14	99.76	4.88	99.56	6.61
50	99.98	1.45	99.90	3.20	99.76	4.94	99.56	6.67
52	99.98	1.51	99.89	3.26	99.75	4.99	99.55	6.73
54	99.98	1.57	99.89	3.31	99.74	5.05	99.54	6.78
56	99.97	1.63	99.89	3.37	99.74	5.11	99.53	6.84
58	99.97	1.69	99.88	3.43	99.73	5.17	99.52	6.90
60	99.97	1.74	99.88	3.49	99.73	5.23	99.51	6.96
c = .75	.75	.01	.75	.02	.75	.03	.75	.05
c = 1.00	1.00	.01	1.00	.03	1.00	.04	1.00	.06
c = 1.25	1.25	.02	1.25	.03	1.25	.05	1.25	.08

HORIZONTAL DISTANCES AND DIFFERENCES OF ELEVATION FOR STADIA MEASUREMENTS.

Minutes.	4°		5°		6°		7°	
	Hor. Dist.	Diff. Elev.	Hor. Dist.	Diff. Elev.	Hor. Dist.	Diff. Elev.	Hor. Dist.	Diff. Elev.
0′.............	99.51	6.96	99.24	8.68	98.91	10.40	98.51	12.10
2	99.51	7.02	99.23	8.74	98.90	10.45	98.50	12.15
4	99.50	7.07	99.22	8.80	98.88	10.51	98.48	12.21
6	99.49	7.13	99.21	8.85	98.87	10.57	98.47	12.26
8	99.48	7.19	99.20	8.91	98.86	10.62	98.46	12.32
10	99.47	7.25	99.19	8.97	98.85	10.68	98.44	12.38
12	99.46	7.30	99.18	9.03	98.83	10.74	98.43	12.43
14	99.46	7.36	99.17	9.08	98.82	10.79	98.41	12.49
16	99.45	7.42	99.16	9.14	98.81	10.85	98.40	12.55
18	99.44	7.48	99.15	9.20	98.80	10.91	98.39	12.60
20	99.43	7.53	99.14	9.25	98.78	10.96	98.37	12.66
22	99.42	7.59	99.13	9.31	98.77	11.02	98.36	12.72
24	99.41	7.65	99.11	9.37	98.76	11.08	98.34	12.77
26	99.40	7.71	99.10	9.43	98.74	11.13	98.33	12.83
28	99.39	7.76	99.09	9.48	98.73	11.19	98.31	12.88
30	99.38	7.82	99.08	9.54	98.72	11.25	98.29	12.94
32	99.38	7.88	99.07	9.60	98.71	11.30	93.28	13.00
34	99.37	7.94	99.06	9.65	98.69	11.36	98.27	13.05
36	99.36	7.99	99.05	9.71	98.68	11.42	98.25	13.11
38	99.35	8.05	99.04	9.77	98.67	11.47	98.24	13.17
40	99.34	8.11	99.03	9.83	98.65	11.53	98.22	13.22
42	99.33	8.17	99.01	9.88	98.64	11.59	98.20	13.28
44	99.32	8.22	99.00	9.94	98.63	11.64	98.19	13.33
46	99.31	8.28	98.99	10.00	98.61	11.70	98.17	13.39
48	99.30	8.34	98.98	10.05	98.60	11.76	98.16	13.45
50	99.29	8.40	98.97	10.11	98.58	11.81	98.14	13.50
52	99.28	8.45	98.96	10.17	98.57	11.87	98.13	13.56
54	99.27	8.51	98.94	10.22	98.56	11.93	98.11	13.61
56	99.26	8.57	98.93	10.28	98.54	11.98	98.10	13.67
58	99.25	8.63	98.92	10.34	98.53	12.04	98.08	13.73
60	99.24	8.68	98.91	10.40	98.51	12.10	98.06	13.78
c = .75	.75	.06	.75	.07	.75	.08	.74	.10
c = 1.00	1.00	.08	.99	.09	.99	.11	.99	.13
c = 1.25	1.25	.10	1.24	.11	1.24	.14	1.24	.16

HORIZONTAL DISTANCES AND DIFFERENCES OF ELEVATION FOR STADIA MEASUREMENTS.

Minutes.	8°		9°		10°		11°	
	Hor. Dist.	Diff. Elev.	Hor. Dist.	Diff. Elev.	Hor. Dist.	Diff. Elev.	Hor. Dist.	Diff. Elev.
0′.............	98.06	13.78	97.55	15.45	96.98	17.10	96.36	18.73
2.............	98.05	13.84	97.53	15.51	96.96	17.16	96.34	18.78
4.............	98.03	13.89	97.52	15.56	96.94	17.21	96.32	18.84
6.............	98.01	13.95	97.50	15.62	96.92	17.26	96.29	18.89
8.............	98.00	14.01	97.48	15.67	96.90	17.32	96.27	18.95
10.............	97.98	14.06	97.46	15.73	96.88	17.37	96.25	19.00
12.............	97.97	14.12	97.44	15.78	96.86	17.43	96.23	19.05
14.............	97.95	14.17	97.43	15.84	96.84	17.48	96.21	19.11
16.............	97.93	14.23	97.41	15.89	96.82	17.54	96.18	19.16
18.............	97.92	14.28	97.39	15.95	96.80	17.59	96.16	19.21
20.............	97.90	14.34	97.37	16.00	96.78	17.65	96.14	19.27
22.............	97.88	14.40	97.35	16.06	96.76	17.70	96.12	19.32
24.............	97.87	14.45	97.33	16.11	96.74	17.76	96.09	19.38
26.............	97.85	14.51	97.31	16.17	96.72	17.81	96.07	19.43
28.............	97.83	14.56	97.29	16.22	96.70	17.86	96.05	19.48
30.............	97.82	14.62	97.28	16.28	96.68	17.92	96.03	19.54
32.............	97.80	14.67	97.26	16.33	96.66	17.97	96.00	19.59
34.............	97.78	14.73	97.24	16.39	96.64	18.03	95.98	19.64
36.............	97.76	14.79	97.22	16.44	96.62	18.08	95.96	19.70
38.............	97.75	14.84	97.20	16.50	96.60	18.14	95.93	19.75
40.............	97.73	14.90	97.18	16.55	96.57	18.19	95.91	19.80
42.............	97.71	14.95	97.16	16.61	96.55	18.24	95.89	19.86
44.............	97.69	15.01	97.14	16.66	96.53	18.30	95.86	19.91
46.............	97.68	15.06	97.12	16.72	96.51	18.35	95.84	19.96
48.............	97.66	15.12	97.10	16.77	96.49	18.41	95.82	20.02
50.............	97.64	15.17	97.08	16.83	96.47	18.46	95.79	20.07
52.............	97.62	15.23	97.06	16.88	96.45	18.51	95.77	20.12
54.............	97.61	15.28	97.04	16.94	96.42	18.57	95.75	20.18
56.............	97.59	15.34	97.02	16.99	96.40	18.62	95.72	20.23
58.............	97.57	15.40	97.00	17.05	96.38	18.68	95.70	20.28
60.............	97.55	15.45	96.98	17.10	96.36	18.73	95.68	20.34
c = .75........	.74	.11	.74	.12	.74	.14	.73	.15
c = 1.00........	.99	.15	.99	.16	.98	.18	.98	.20
c = 1.25........	1.23	.18	1.23	.21	1.23	.23	1.22	.25

HORIZONTAL DISTANCES AND DIFFERENCES OF ELEVATION FOR STADIA MEASUREMENTS.

Minutes.	12°		13°		14°		15°	
	Hor. Dist.	Diff. Elev.	Hor. Dist.	Diff. Elev.	Hor. Dist.	Diff. Elev.	Hor. Dist.	Diff. Elev.
0′.............	95.68	20.34	94.94	21.92	94.15	23.47	93.30	25.00
2	95.65	20.39	94.91	21.97	94.12	23.52	93.27	25.05
4	95.63	20.44	94.89	22.02	94.09	23.58	93.24	25.10
6	95.61	20.50	94.86	22.08	94.07	23.63	93.21	25.15
8	95.58	20.55	94.84	22.13	94.04	23.68	93.18	25.20
10	95.56	20.60	94.81	22.18	94.01	23.73	93.16	25.25
12	95.53	20.66	94.79	22.23	93.98	23.78	93.13	25.30
14	95.51	20.71	94.76	22.28	93.95	23.83	93.10	25.35
16	95.49	20.76	94.73	22.34	93.93	23.88	93.07	25.40
18	95.46	20.81	94.71	22.39	93.90	23.93	93.04	25.45
20	95.44	20.87	94.68	22.44	93.87	23.99	93.01	25.50
22	95.41	20.92	94.66	22.49	93.84	24.04	92.98	25.55
24	95.39	20.97	94.63	22.54	93.81	24.09	92.95	25.60
26	95.36	21.03	94.60	22.60	93.79	24.14	92.92	25.65
28	95.34	21.08	94.58	22.65	93.76	24.19	92.89	25.70
30	95.32	21.13	94.55	22.70	93.73	24.24	92.86	25.75
32	95.29	21.18	94.52	22.75	93.70	24.29	92.83	25.80
34	95.27	21.24	94.50	22.80	93.67	24.34	92.80	25.85
36	95.24	21.29	94.47	22.85	93.65	24.39	92.77	25.90
38	95.22	21.34	94.44	22.91	93.62	24.44	92.74	25.95
40	95.19	21.39	94.42	22.96	93.59	24.49	92.71	26.00
42	95.17	21.45	94.39	23.01	93.56	24.55	92.68	26.05
44	95.14	21.50	94.36	23.06	93.53	24.60	92.65	26.10
46	95.12	21.55	94.34	23.11	93.50	24.65	92.62	26.15
48	95.09	21.60	94.31	23.16	93.47	24.70	92.59	26.20
50	95.07	21.66	94.28	23.22	93.45	24.75	92.56	26.25
52	95.04	21.71	94.26	23.27	93.42	24.80	92.53	26.30
54	95.02	21.76	94.23	23.32	93.39	24.85	92.49	26.35
56	94.99	21.81	94.20	23.37	93.36	24.90	92.46	26.40
58	94.97	21.87	94.17	23.42	93.33	24.95	92.43	26.45
60	94.94	21.92	94.15	23.47	93.30	25.00	92.40	26.50
c = .75	.73	.16	.73	.17	.73	.19	.72	.20
c = 1.00	.98	.22	.97	.23	.97	.25	.96	.27
c = 1.25	1.22	.27	1.21	.29	1.21	.31	1.20	.34

HORIZONTAL DISTANCES AND DIFFERENCES OF ELEVATION FOR STADIA MEASUREMENTS.

Minutes.	16°		17°		18°		19°	
	Hor. Dist.	Diff. Elev.	Hor. Dist.	Diff. Elev.	Hor. Dist.	Diff. Elev.	Hor. Dist.	Diff. Elev.
0′	92.40	26.50	91.45	27.96	90.45	29.39	89.40	30.78
2	92.37	26.55	91.42	28.01	90.42	29.44	89.36	30.83
4	92.34	26.59	91.39	28.06	90.38	29.48	89.33	30.87
6	92.31	26.64	91.35	28.10	90.35	29.53	89.29	30.92
8	92.28	26.69	91.32	28.15	90.31	29.58	89.26	30.97
10	92.25	26.74	91.29	28.20	90.28	29.62	89.22	31.01
12	92.22	26.79	91.26	28.25	90.24	29.67	89.18	31.06
14	92.19	26.84	91.22	28.30	90.21	29.72	89.15	31.10
16	92.15	26.89	91.19	28.34	90.18	29.76	89.11	31.15
18	92.12	26.94	91.16	28.39	90.14	29.81	89.08	31.19
20	92.09	26.99	91.12	28.44	90.11	29.86	89.04	31.24
22	92.06	27.04	91.09	28.49	90.07	29.90	89.00	31.28
24	92.03	27.09	91.06	28.54	90.04	29.95	88.96	31.33
26	92.00	27.13	91.02	28.58	90.00	30.00	88.93	31.38
28	91.97	27.18	90.99	28.63	89.97	30.04	88.89	31.42
30	91.93	27.23	90.96	28.68	89.93	30.09	88.86	31.47
32	91.90	27.28	90.92	28.73	89.90	30.14	88.82	31.51
34	91.87	27.33	90.89	28.77	89.86	30.19	88.78	31.56
36	91.84	27.38	90.86	28.82	89.83	30.23	88.75	31.60
38	91.81	27.43	90.82	28.87	89.79	30.28	88.71	31.65
40	91.77	27.48	90.79	28.92	89.76	30.32	88.67	31.69
42	91.74	27.52	90.76	28.96	89.72	30.37	88.64	31.74
44	91.71	27.57	90.72	29.01	89.69	30.41	88.60	31.78
46	91.68	27.62	90.69	29.06	89.65	30.46	88.56	31.83
48	91.65	27.67	90.66	29.11	89.61	30.51	88.53	31.87
50	91.61	27.72	90.62	29.15	89.58	30.55	88.49	31.92
52	91.58	27.77	90.59	29.20	89.54	30.60	88.45	31.96
54	91.55	27.81	90.55	29.25	89.51	30.65	88.41	32.01
56	91.52	27.86	90.52	29.30	89.47	30.69	88.38	32.05
58	91.48	27.91	90.48	29.34	89.44	30.74	88.34	32.09
60	91.45	27.96	90.45	29.39	89.40	30.78	88.30	32.14
c = .75	.72	.21	.72	.23	.71	.24	.71	.25
c = 1.00	.86	.28	.95	.30	.95	.32	.94	.33
c = 1.25	1.20	.35	1.19	.38	1.19	.40	1.18	.42

HORIZONTAL DISTANCES AND DIFFERENCES
OF ELEVATION FOR STADIA MEASUREMENTS.

Minutes.	20°		21°		22°		23°	
	Hor. Dist.	Diff. Elev.	Hor. Dist.	Diff. Elev.	Hor. Dist.	Diff. Elev.	Hor. Dist.	Diff. Elev.
0′	88.30	32.14	87.16	33.46	85.97	34.73	84.73	35.97
2	88.26	32.18	87.12	33.50	85.93	34.77	84.69	36.01
4	88.23	32.23	87.08	33.54	85.89	34.82	84.65	36.05
6	88.19	32.27	87.04	33.59	85.85	34.86	84.61	36.09
8	88.15	32.32	87.00	33.63	85.80	34.90	84.57	36.13
10	88.11	32.36	86.96	33.67	85.76	34.94	84.52	36.17
12	88.08	32.41	86.92	33.72	85.72	34.98	84.48	36.21
14	88.04	32.45	86.88	33.76	85.68	35.02	84.44	36.25
16	88.00	32.49	86.84	33.80	85.64	35.07	84.40	36.29
18	87.96	32.54	86.80	33.84	85.60	35.11	84.35	36.33
20	87.93	32.58	86.77	33.89	85.56	35.15	84.31	36.37
22	87.89	32.63	86.73	33.93	85.52	35.19	84.27	36.41
24	87.85	32.67	86.69	33.97	85.48	35.23	84.23	36.45
26	87.81	32.72	86.65	34.01	85.44	35.27	84.18	36.49
28	87.77	32.76	86.61	34.06	85.40	35.31	84.14	36.53
30	87.74	32.80	86.57	34.10	85.36	35.36	84.10	36.57
32	87.70	32.85	86.53	34.14	85.31	35.40	84.06	36.61
34	87.66	32.89	86.49	34.18	85.27	35.44	84.01	36.65
36	87.62	32.93	86.45	34.23	85.23	35.48	83.97	36.69
38	87.58	32.98	86.41	34.27	85.19	35.52	83.93	36.73
40	87.54	33.02	86.37	34.31	85.15	35.56	83.89	36.77
42	87.51	33.07	86.33	34.35	85.11	35.60	83.84	36.80
44	87.47	33.11	86.29	34.40	85.07	35.64	83.80	36.84
46	87.43	33.15	86.25	34.44	85.02	35.68	83.76	36.88
48	87.39	33.20	86.21	34.48	84.98	35.72	83.72	36.92
50	87.35	33.24	86.17	34.52	84.94	35.76	83.67	36.96
52	87.31	33.28	86.13	34.57	84.90	35.80	83.63	37.00
54	87.27	33.33	86.09	34.61	84.86	35.85	83.59	37.04
56	87.24	33.37	86.05	34.65	84.82	35.89	83.54	37.08
58	87.20	33.41	86.01	34.69	84.77	35.93	83.50	37.12
60	87.16	33.46	85.97	34.73	84.73	35.97	83.46	37.16
c = .75	.70	.26	.70	.27	.69	.29	.69	.30
c = 1.00	.94	.35	.93	.37	.92	.38	.92	.40
c = 1.25	1.17	.44	1.16	.46	1.15	.48	1.15	.50

HORIZONTAL DISTANCES AND DIFFERENCES
OF ELEVATION FOR STADIA MEASUREMENTS.

Minutes.	24°		25°		26°		27°	
	Hor. Dist.	Diff. Elev.	Hor. Dist.	Diff. Elev.	Hor. Dist.	Diff. Elev.	Hor. Dist.	Diff. Elev.
0′............	83.46	37.16	82.14	38.30	80.78	39.40	79.39	40.45
2............	83.41	37.20	82.09	38.34	80.74	39.44	79.34	40.49
4............	83.37	37.23	82.05	38.38	80.69	39.47	79.30	40.52
6............	83.33	37.27	82.01	38.41	80.65	39.51	79.25	40.55
8............	83.28	37.31	81.96	38.45	80.60	39.54	79.20	40.59
10	83.24	37.35	81.92	38.49	80.55	39.58	79.15	40.62
12	83.20	37.39	81.87	38.53	80.51	39.61	79.11	40.66
14	83.15	37.43	81.83	38.56	80.46	39.65	79.06	40.69
16	83.11	37.47	81.78	38.60	80.41	39.69	79.01	40.72
18	83.07	37.51	81.74	38.64	80.37	39.72	78.96	40.76
20	83.02	37.54	81.69	38.67	80.32	39.76	78.92	40.79
22	82.98	37.58	81.65	38.71	80.28	39.79	78.87	40.82
24	82.93	37.62	81.60	38.75	80.23	39.83	78.82	40.86
26	82.89	37.66	81.56	38.78	80.18	39.86	78.77	40.89
28	82.85	37.70	81.51	38.82	80.14	39.90	78.73	40.92
30	82.80	37.74	81.47	38.86	80.09	39.93	78.68	40.96
32	82.76	37.77	81.42	38.89	80.04	39.97	78.63	40.99
34	82.72	37.81	81.38	38.93	80.00	40.00	78.58	41.02
36	82.67	37.85	81.33	38.97	79.95	40.04	78.54	41.06
38	82.63	37.89	81.28	39.00	79.90	40.07	78.49	41.09
40	82.58	37.93	81.24	39.04	79.86	40.11	78.44	41.12
42	82.54	37.96	81.19	39.08	79.81	40.14	78.39	41.16
44	82.49	38.00	81.15	39.11	79.76	40.18	78.34	41.19
46	82.45	38.04	81.10	39.15	79.72	40.21	78.30	41.22
48	82.41	38.08	81.06	39.18	79.67	40.24	78.25	41.26
50	82.36	38.11	81.01	39.22	79.62	40.28	78.20	41.29
52	82.32	38.15	80.97	39.26	79.58	40.31	78.15	41.32
54	82.27	38.19	80.92	39.29	79.53	40.35	78.10	41.35
56	82.23	38.23	80.87	39.33	79.48	40.38	78.06	41.39
58	82.18	38.26	80.83	39.36	79.44	40.42	78.01	41.42
60	82.14	38.30	80.78	39.40	79.39	40.45	77.96	41.45
c = .75	.68	.31	.68	.32	.67	.33	.66	.35
c = 1.00	.91	.41	.90	.43	.89	.45	.89	.46
c = 1.25	1.14	.52	1.13	.54	1.12	.56	1.11	.58

HORIZONTAL DISTANCES AND DIFFERENCES OF ELEVATION FOR STADIA MEASUREMENTS.

Minutes.	28°		29°		30°	
	Hor. Dist.	Diff. Elev.	Hor. Dist.	Diff. Elev.	Hor. Dist.	Diff. Elev.
0′.............	77.96	41.45	76.50	42.40	75.00	43.30
2	77.91	41.48	76.45	42.43	74.95	43.33
4	77.86	41.52	76.40	42.46	74.90	43.36
6	77.81	41.55	76.35	42.49	74.85	43.39
8	77.77	41.58	76.30	42.53	74.80	43.42
10	77.72	41.61	76.25	42.56	74.75	43.45
12	77.67	41.65	76.20	42.59	74.70	43.47
14	77.62	41.68	76.15	42.62	74.65	43.50
16	77.57	41.71	76.10	42.65	74.60	43.53
18	77.52	41.74	76.05	42.68	74.55	43.56
20	77.48	41.77	76.00	42.71	74.49	43.59
22	77.42	41.81	75.95	42.74	74.44	43.62
24	77.38	41.84	75.90	42.77	74.39	43.65
26	77.33	41.87	75.85	42.80	74.34	43.67
28 ...:.........	77.28	41.90	75.80	42.83	74.29	43.70
30	77.23	41.93	75.75	42.86	74.24	43.73
32	77.18	41.97	75.70	42.89	74.19	43.76
34	77.13	42.00	75.65	42.92	74.14	43.79
36	77.09	42.03	75.60	42.95	74.09	43.82
38	77.04	42.06	75.55	42.98	74.04	43.84
40	76.99	42.09	75.50	43.01	73.99	43.87
42 ,............	76.94	42.12	75.45	43.04	73.93	43.90
44	76.89	42.15	75.40	43.07	73.88	43.93
46	76.84	42.19	75.35	43.10	73.83	43.95
48	76.79	42.22	75.30	43.13	73.78	43.98
50	76.74	42.25	75.25	43.16	73.73	44.01
52	76.69	42.28	75.20	43.18	73.68	44.04
54	76.64	42.31	75.15	43.21	73.63	44.07
56	76.59	42.34	75.10	43.24	73.58	44.09
58	76.55	42.37	75.05	43.27	73.52	44.12
60	76.50	42.40	75.00	43.30	73.47	44.15
c = .75	.66	.36	.65	.37	.65	.38
c = 1.00	.88	.48	.87	.49	.86	.51
c = 1.25	1.10	.60	1.09	.62	1.08	.64

TABLE OF

RADII AND CHORD AND TANGENT DEFLECTIONS.

The formulas used in the computation of the following tables are as follows:

For Radii, $R = \dfrac{50}{\sin D}$.   (89.)   Art. **1249.**

For Chord Deflections,

$$d = \frac{c^2}{R}.$$ (92.)   Art. **1255.**

For Tangent Deflections,

$$\tan \text{deflection} = \frac{c^2}{2R}.$$ (93.)   Art. **1255.**

TABLE OF RADII AND DEFLECTIONS.

Degree	Radii	Chord Deflection	Tangent Deflection	Degree	Radii	Chord Deflection	Tangent Deflection	Degree	Radii	Chord Deflection	Tangent Deflection
0 5	68754.94	.145	.073	5 15	1091.73	9.160	4.580	10 50	529.67	18.880	9.440
10	34377.48	.291	.145	20	1074.68	9.305	4.653	11 0	521.67	19.169	9.585
15	22918.33	.436	.218	25	1058.16	9.450	4.725	10	513.91	19.459	9.729
20	17188.76	.582	.291	30	1042.14	9.596	4.798	20	506.38	19.748	9.874
25	13751.02	.727	.364	35	1026.60	9.741	4.870	30	499.06	20.038	10.019
30	11459.19	.873	.436	40	1011.51	9.886	4.943	40	491.96	20.327	10.164
35	9822.18	1.018	.509	45	996.87	10.031	5.016	50	485.05	20.616	10.308
40	8594.41	1.164	.582	50	982.64	10.177	5.088	12 0	478.34	20.906	10.453
45	7639.49	1.309	.654	55	968.81	10.322	5.161	10	471.81	21.195	10.597
50	6875.55	1.454	.727	6 0	955.37	10.467	5.234	20	465.46	21.484	10.742
55	6250.51	1.600	.800	5	942.29	10.612	5.306	30	459.28	21.773	10.887
1 0	5729.65	1.745	.873	10	929.57	10.758	5.379	40	453.26	22.063	11.031
5	5288.92	1.891	.945	15	917.19	10.903	5.451	50	447.40	22.352	11.176
10	4911.15	2.036	1.018	20	905.13	11.048	5.524	13 0	441.68	22.641	11.320
15	4583.75	2.182	1.091	25	893.39	11.193	5.597	10	436.12	22.930	11.465
20	4297.28	2.327	1.164	30	881.95	11.339	5.669	20	430.69	23.219	11.609
25	4044.51	2.472	1.236	35	870.79	11.484	5.742	30	425.40	23.507	11.754
30	3819.83	2.618	1.309	40	859.92	11.629	5.814	40	420.23	23.796	11.898
35	3618.80	2.763	1.382	45	849.32	11.774	5.887	50	415.19	24.085	12.043
40	3437.87	2.909	1.454	50	838.97	11.910	5.960	14 0	410.28	24.374	12.187
45	3274.17	3.054	1.527	55	828.88	12.065	6.032	10	405.47	24.663	12.331
50	3125.36	3.200	1.600	7 0	819.02	12.210	6.105	20	400.78	24.951	12.476
55	2989.48	3.345	1.673	5	809.40	12.355	6.177	30	396.20	25.240	12.620
2 0	2864.93	3.490	1.745	10	800.00	12.500	6.250	40	391.72	25.528	12.764
5	2750.35	3.636	1.818	15	790.81	12.645	6.323	50	387.34	25.817	12.908
10	2644.58	3.781	1.891	20	781.84	12.790	6.395	15 0	383.06	26.105	13.053
15	2546.64	3.927	1.963	25	773.07	12.936	6.468	10	378.88	26.394	13.197
20	2455.70	4.072	2.036	30	764.49	13.081	6.540	20	374.79	26.682	13.341
25	2371.04	4.218	2.109	35	756.10	13.226	6.613	30	370.78	26.970	13.485
30	2292.01	4.363	2.181	40	747.89	13.371	6.685	40	366.86	27.258	13.629
35	2218.09	4.508	2.254	45	739.86	13.516	6.758	50	363.02	27.547	13.773
40	2148.79	4.654	2.327	50	732.01	13.661	6.831	16 0	359.26	27.835	13.917
45	2083.68	4.799	2.400	55	724.31	13.806	6.903	10	355.59	28.123	14.061
50	2022.41	4.945	2.472	8 0	716.78	13.951	6.976	20	351.98	28.411	14.205
55	1964.64	5.090	2.545	5	709.40	14.096	7.048	30	348.45	28.699	14.349
3 0	1910.08	5.235	2.618	10	702.18	14.241	7.121	40	344.99	28.986	14.493
5	1858.47	5.381	2.690	15	695.09	14.387	7.193	50	341.60	29.274	14.637
10	1809.57	5.526	2.763	20	688.16	14.532	7.266	17 0	338.27	29.562	14.781
15	1763.18	5.672	2.836	25	681.35	14.677	7.338	10	335.01	29.850	14.925
20	1719.12	5.817	2.908	30	674.69	14.822	7.411	20	331.82	30.137	15.069
25	1677.20	5.962	2.981	35	668.15	14.967	7.483	30	328.68	30.425	15.212
30	1637.28	6.108	3.054	40	661.74	15.112	7.556	40	325.60	30.712	15.356
35	1599.21	6.253	3.127	45	655.45	15.257	7.628	50	322.59	31.000	15.500
40	1562.88	6.398	3.199	50	649.27	15.402	7.701	18 0	319.62	31.287	15.643
45	1528.16	6.544	3.272	55	643.22	15.547	7.773	10	316.71	31.574	15.787
50	1494.95	6.689	3.345	9 0	637.27	15.692	7.846	20	313.86	31.861	15.931
55	1463.16	6.835	3.417	5	631.44	15.837	7.918	30	311.06	32.149	16.074
4 0	1432.69	6.980	3.490	10	625.71	15.982	7.991	40	308.30	32.436	16.218
5	1403.46	7.125	3.563	15	620.09	16.127	8.063	50	305.60	32.723	16.361
10	1375.40	7.271	3.635	20	614.56	16.272	8.136	19 0	302.94	33.010	16.505
15	1348.45	7.416	3.708	25	609.14	16.417	8.208	10	300.33	33.296	16.648
20	1322.53	7.561	3.781	30	603.80	16.562	8.281	20	297.77	33.583	16.792
25	1297.58	7.707	3.853	35	598.57	16.707	8.353	30	295.25	33.870	16.935
30	1273.57	7.852	3.926	40	593.42	16.852	8.426	40	292.77	34.157	17.078
35	1250.42	7.997	3.999	45	588.36	16.996	8.498	50	290.33	34.443	17.222
40	1228.11	8.143	4.071	50	583.38	17.141	8.571	20 0	287.94	34.730	17.365
45	1206.57	8.288	4.144	55	578.49	17.286	8.643				
50	1185.78	8.433	4.217	10 0	573.69	17.431	8.716				
55	1165.70	8.579	4.289	10	564.31	17.721	8.860				
5 0	1146.28	8.724	4.362	20	555.23	18.011	9.005				
5	1127.50	8.869	4.435	30	546.44	18.300	9.150				
10	1109.33	9.014	4.507	40	537.92	18.590	9.295				

MOMENTS OF INERTIA.

Form of Section.	Dotted Line Shows Position of Neutral Axis.	A	I	c
1. Rectangle ...		bd	$\frac{1}{12}bd^2$	$\frac{1}{2}d$
2. Square ..		d^2	$\frac{1}{12}d^4$	$\frac{1}{2}d$
3. Axis through Diagonal		d^2	$\frac{1}{12}d^4$	$.707d$
4. Hollow Square		$d^2 - d_1^2$	$\frac{1}{12}(d^4 - d_1^4)$	$\frac{1}{2}d$
5. Hollow Rectangle, I or Channel Iron		$bd - b_1d_1$	$\frac{1}{12}(bd^2 - b_1d_1^2)$	$\frac{1}{2}d$
6. Triangle		$\frac{1}{2}bd$	$\frac{1}{36}bd^2$	$\frac{2}{3}d$
7. Cross....		$td + t_1b$	$\frac{1}{12}(td^2 + bt_1^2)$	$\frac{1}{2}d$
8. Angle Iron.....		$bd - b_1d_1$	$\dfrac{(bd^2 - b_1d_1^2)^2 - 4bdb_1d_1(d-d_1)^2}{12(bd - b_1d_1)}$	$\dfrac{d}{2} + \dfrac{b_1d_1}{2}\left(\dfrac{d-d_1}{bd - b_1d_1}\right)$
9. Circle....		$\frac{\pi}{4}d^2$	$\dfrac{\pi d^4}{64}$	$\frac{1}{2}d$
10. Hollow Circle...		$\frac{\pi}{4}(d^2 - d_1^2)$	$\dfrac{\pi(d^4 - d_1^4)}{64}$	$\frac{1}{2}d$
11. Ellipse..		$\frac{\pi}{4}bd$	$\dfrac{\pi bd^2}{64}$	$\frac{1}{2}d$
12. Hollow Ellipse..		$\frac{\pi}{4}(bd - b_1d_1)$	$\dfrac{\pi(bd^2 - b_1d_1^2)}{64}$	$\frac{1}{2}d$

BENDING MOMENTS AND DEFLECTIONS.

Manner of Supporting Beams.	Maximum Bending Moment, M.	Maximum Deflection, s.	Remarks.
1.	Wl	$\dfrac{1}{3}\dfrac{Wl^2}{EI}$	Cantilever, load at free end.
2.	$W_1 l_1 + W_2 l_2$		Cantilever, more than one load.
3.	$\dfrac{wl^2}{2}$	$\dfrac{1}{8}\dfrac{Wl^2}{EI}$	Cantilever, uniform load w lb. per unit of length.
4.	$\dfrac{wl^2}{2}+Wl$	$\dfrac{1}{3}\dfrac{Wl^3}{EI}+\dfrac{1}{8}\dfrac{Wl^3}{EI}$	Cantilever, load partly uniform, partly concentrated.
5.	$\dfrac{Wl}{4}$	$\dfrac{1}{48}\dfrac{Wl^3}{EI}$	Simple beam, load at middle.
6.	$W\dfrac{l_1 l_2}{l}$		Simple beam, load at some other point than the middle.
7.	$\dfrac{wl^2}{8}$	$\dfrac{5}{384}\dfrac{Wl^3}{EI}$	Simple beam, uniformly loaded.
8.	$\dfrac{3}{16}Wl$	$.0182\dfrac{Wl^3}{EI}$	One end fixed, other end supported, load in the middle.
9.	$\dfrac{wl^2}{8}$	$.0054\dfrac{Wl^3}{EI}$	One end fixed, other end supported, uniformly loaded.
10.	$\dfrac{Wl}{8}$	$\dfrac{1}{192}\dfrac{Wl^3}{EI}$	Both ends fixed, load in the middle.
11.	$\dfrac{wl^2}{12}$	$\dfrac{1}{384}\dfrac{Wl^3}{EI}$	Both ends fixed, uniformly loaded.

SPECIFIC GRAVITIES AND WEIGHTS PER CUBIC FOOT.

METALS.

Substance.	Specific Gravity.	Weight per Cubic Foot in Pounds.
Osmium	23.00	1,437.5
Platinum	21.50	1,343.8
Gold	19.50	1,218.8
Mercury	13.60	850.0
Lead (cast)	11.35	709.4
Silver	10.50	656.3
Copper (cast)	8.79	549.4
Brass	8.38	523.8
Wrought Iron	7.68	480.0
Cast Iron	7.21	450.0
Steel	7.84	490.0
Tin (cast)	7.29	455.6
Zinc (cast)	6.86	428.8
Antimony	6.71	419.4
Aluminum	2.50	156.3

WOODS.

Substance.	Specific Gravity.	Weight per Cubic Foot in Pounds.
Ash	.845	52.80
Beech	.852	53.25
Cedar	.561	35.06
Cork	.240	15.00
Ebony (American)	1.331	83.19
Lignum-vitæ	1.333	83.30
Maple	.750	46.88
Oak (old)	1.170	73.10
Spruce	.500	31.25
Pine (yellow)	.660	41.20
Pine (white)	.554	34.60
Walnut	.671	41.90

LIQUIDS.

Substance.	Specific Gravity.	Weight per Cubic Foot in Pounds.
Acetic Acid	1.062	66.4
Nitric Acid	1.217	76.1
Sulphuric Acid	1.841	115.1
Muriatic Acid	1.200	75.0
Alcohol	.800	50.0
Turpentine........................	.870	54.4
Sea Water (ordinary)	1.026	64.1
Milk...............................	1.032	64.5

GASES.

At 32° F., and under a Pressure of One Atmosphere.

Substance.	Specific Gravity.	Weight per Cubic Foot in Pounds.
Atmospheric Air....................	1.0000	.08073
Carbonic Acid.....................	1.5290	.12344
Carbonic Oxide....................	.9674	.07810
Chlorine	2.4400	.19700
Oxygen	1.1056	.08925
Nitrogen..........................	.9736	.07860
Smoke (bituminous coal)...........	.1020	.00815
Smoke (wood).....................	.0900	.00727
*Steam at 212° F.................	.4700	.03790
Hydrogen.........................	.0692	.00559

* The specific gravity of steam at any temperature and pressure compared with air at the same temperature and pressure is 0.022.

MISCELLANEOUS.

Substance.	Specific Gravity.	Weight per Cubic Foot in Pounds.
Emery	4.00	250
Glass (average)	2.80	175
Chalk	2.78	174
Granite	2.65	166
Marble	2.70	169
Stone (common)	2.52	158
Salt (common)	2.13	133
Soil (common)	1.98	124
Clay	1.93	121
Brick	1.90	118
Plaster Paris (average)	2.00	125
Sand	1.80	113

COEFFICIENTS FOR FLOW OF WATER.

DISCHARGE OF STANDARD ORIFICES.

COEFFICIENTS FOR CIRCULAR VERTICAL ORIFICES.

Head h in Feet.	Diameter of Orifice in Feet.						
	0.02	0.04	0.07	0.10	0.20	0.60	1.00
0.4		0.637	0.624	0.618			
0.6	0.655	.630	.618	.613	0.601	0.593	
0.8	.648	.626	.615	.610	.601	.594	0.590
1.0	.644	.623	.612	.608	.600	.595	.591
1.5	.637	.618	.608	.605	.600	.596	.593
2.0	.632	.614	.607	.604	.599	.597	.595
2.5	.629	.612	.605	.603	.599	.598	.596
3.0	.627	.611	.604	.603	.599	.598	.597
4.0	.623	.609	.603	.602	.599	.597	.596
6.0	.618	.607	.602	.600	.598	.597	.596
8.0	.614	.605	.601	.600	.598	.596	.596
10.0	.611	.603	.599	.598	.597	.596	.595
20.0	.601	.599	.597	.596	.596	.596	.594
50.0	.596	.595	.594	.594	.594	.594	.593
100.0	.593	.592	.592	.592	.592	.592	.592

COEFFICIENTS FOR SQUARE VERTICAL ORIFICES.

Head h in Feet.	Side of the Square in Feet.						
	0.02	0.04	0.07	0.10	0.20	0.60	1.00
0.4		0.643	0.628	0.621			
0.6	0.660	.636	.623	.617	0.605	0.598	
0.8	.652	.631	.620	.615	.605	.600	0.597
1.0	.648	.628	.618	.613	.605	.601	.599
1.5	.641	.622	.614	.610	.605	.602	.601
2.0	.637	.619	.612	.608	.605	.604	.602
2.5	.634	.617	.610	.607	.605	.604	.602
3.0	.632	.616	.609	.607	.605	.604	.603
4.0	.628	.614	.608	.606	.605	.603	.602
6.0	.623	.612	.607	.605	.604	.603	.602
8.0	.619	.610	.606	.605	.604	.603	.602
10.0	.616	.608	.605	.604	.603	.602	.601
20.0	.606	.604	.602	.602	.602	.601	.600
50.0	.602	.601	.601	.600	.600	.599	.599
100.0	.599	.598	.598	.598	.598	.598	.598

COEFFICIENTS FOR RECTANGULAR ORIFICES 1 FOOT WIDE.

Head h on Center of Orifice in Feet.	Depth of Orifice in Feet.						
	0.125	0.25	0.50	0.75	1.00	1.50	2.00
0.4	0.634	0.633	0.622				
0.6	.633	.633	.619	0.614			
0.8	.633	.633	.618	.612	0.608		
1.0	.632	.632	.618	.612	.606	0.626	
1.5	.630	.631	.618	.611	.605	.626	0.628
2.0	.629	.630	.617	.611	.605	.624	.630
2.5	.628	.628	.616	.611	.605	.616	.627
3.0	.627	.627	.615	.610	.605	.614	.619
4.0	.624	.624	.614	.609	.605	.612	.616
6.0	.615	.615	.609	.604	.602	.606	.610
8.0	.609	.607	.603	.602	.601	.602	.604
10.0	.606	.603	.601	.601	.601	.601	.602
20.0				.601	.601	.601	.602

DISCHARGE OF WEIRS.

COEFFICIENTS FOR WEIRS WITH END CONTRACTIONS.

Effective Head in Feet.	Length of Weir in Feet.						
	0.66	1	2	3	5	10	19
0.1	0.632	0.639	0.646	0.652	0.653	0.655	0.656
0.15	.619	.625	.634	.638	.640	.641	.642
0.20	.611	.618	.626	.630	.631	.633	.634
0.25	.605	.612	.621	.624	.626	.628	.629
0.30	.601	.608	.616	.619	.621	.624	.625
0.40	.595	.601	.609	.613	.615	.618	.620
0.50	.590	.596	.605	.608	.611	.615	.617
0.60	.587	.593	.601	.605	.608	.613	.615
0.70		.590	.598	.603	.606	.612	.614
0.80			.595	.600	.604	.611	.613
0.90			.592	.598	.603	.609	.612
1.00			.590	.595	.601	.608	.611
1.2			.585	.591	.597	.605	.610
1.4			.580	.587	.594	.602	.609
1.6				.582	.591	.600	.607

NOTE.—The head given is the effective head, $H + \frac{1}{3}h$. When the velocity of approach is small, h is neglected.

COEFFICIENTS FOR WEIRS WITHOUT END CONTRACTIONS.

Effective Head in Feet.	Length of Weir in Feet.						
	19	10	7	5	4	3	2
0.10	0.657	0.658	0.658	0.659			
0.15	.643	.644	.645	.645	0.647	0.649	0.652
0.20	.635	.637	.637	.638	.641	.642	.645
0.25	.630	.632	.633	.634	.636	.638	.641
0.30	.626	.628	.629	.631	.633	.636	.639
0.40	.621	.623	.625	.628	.630	.633	.636
0.50	.619	.621	.624	.627	.630	.633	.637
0.60	.618	.620	.623	.627	.630	.634	.638
0.70	.618	.620	.624	.628	.631	.635	.640
0.80	.618	.621	.625	.629	.633	.637	.643
0.90	.619	.622	.627	.631	.635	.639	.645
1.00	.619	.624	.628	.633	.637	.641	.648
1.2	.620	.626	.632	.636	.641	.646	
1.4	.622	.629	.634	.640	.644		
1.6	.623	.631	.637	.642	.647		

NOTE.—The head given is the effective head, $H + \frac{1}{3}h$. When the velocity of approach is small, h may be neglected.

FLOW THROUGH PIPES.

VALUES OF THE COEFFICIENT OF FRICTION *f* FOR SMOOTH CAST OR WROUGHT IRON PIPES, EITHER WITH OR WITHOUT COAL-TAR COATING.

Diameter.		Velocity Feet per Second.									
Inches.	Feet.	1	2	3	4	5	6	8	10	12	15
½	.04166	.0505	.0435	.0382	.0355	.0340	.0325	.0315	.0305	.0303	.030
¾	.06250	.0440	.0370	.0335	.0310	.0298	.0287	.0277	.0267	.0262	.0255
1	.08333	.0400	.0335	.0305	.0285	.0276	.0267	.0257	.0248	.0243	.0235
1¼	.125	.0365	.0305	.0285	.0263	.0259	.0255	.0242	.0230	.0226	.0220
1½	.14583	.0350	.0292	.0275	.0257	.0253	.0250	.0238	.0227	.0221	.0215
2	.16666	.0340	.0283	.0268	.0253	.0250	.0247	.0236	.0225	.0219	.0210
3	.25	.0320	.0275	.0260	.0245	.0241	.0237	.0226	.0215	.0210	.0203
4	.33333	.0305	.0268	.0256	.0240	.0235	.0230	.0219	.0208	.0204	.0197
6	.5	.0280	.0260	.0250	.0233	.0226	.0220	.0210	.0200	.0196	.0190
8	.66666	.0265	.0253	.0243	.0227	.0220	.0213	.0203	.0193	.0189	.0183
10	.83333	.0255	.0247	.0237	.0220	.0213	.0207	.0197	.0187	.0183	.0177
12	1.0000	.0248	.0240	.0230	.0213	.0207	.0200	.0190	.0180	.0176	.0170
16	1.33333	.0238	.0227	.0217	.0200	.0193	.0187	.0177	.0167	.0163	.0157
18	1.5000	.0230	.0220	.0210	.0193	.0187	.0180	.0170	.0160	.0156	.0150
20	1.66666	.0223	.0212	.0203	.0187	.0180	.0173	.0163	.0153	.0149	.0143
24	2.000	.0210	.0200	.0190	.0170	.0165	.0160	.0150	.0140	.0136	.0130
30	2.50	.0200	.0190	.0180	.0160	.0155	.0150	.0140	.0130	.0126	.0120
36	3.000	.0190	.0180	.0160	.0150	.0145	.0140	.0135	.0130	.0126	.0120
42	3.500	.0180	.0170	.0160	.0140	.0135	.0130	.0125	.0120	.0116	
48	4.00	.0170	.0160	.0150	.0130	.0125	.0120	.0115	.0110		
60	5.00	.0160	.0150	.0140	.0130	.0125	.0120				
72	6.00	.0150	.0140	.0130	.0120	.0115	.0110				

COEFFICIENTS FOR ANGULAR BENDS.

a° = angle of bend in degrees.

$a° =$	10°	20°	40°	60°	80°	90°	100°	110°	120°	130°	140°	150°
$c =$	.017	.046	.139	.364	.74	.984	1.26	1.56	1.86	2.16	2.43	2.81

COEFFICIENTS FOR CIRCULAR BENDS.

r = radius of pipe.   R = radius of bend.

$\frac{r}{R} =$	.1	.2	.3	.4	.5	.6	.7	.8	.9	1.0
$c' =$	.131	.138	.158	.206	.294	.440	.661	.977	1.408	1.978

COEFFICIENTS FOR DARCY'S FORMULA.

Diameter of Pipe in Inches.	Coefficients for Rough Pipes.	Coefficients for Smooth Pipes.
3	0.00080	0.00040
4	.00076	.00038
6	.00072	.00036
8	.00068	.00034
10	.00066	.00033
12	.00066	.00033
14	.00065	.00033
16	.00064	.00032
24	.00064	.00032
30	.00063	.00032
36	.00062	.00031
48	.00062	.00031

THE PROPERTIES OF SATURATED STEAM.

Pressure above Vacuum in Pounds per Square Inch.	Temperature, Fahrenheit Degrees.	Quantities of Heat in British Thermal Units.			Weight of a Cubic Foot of Steam in Pounds.	Volume.	
		Required to Raise Temperature of the Water from 32° to t'.	Total Latent Heat at Pressure P.	Total Heat above 32°.		Of a Pound of Steam in Cubic Feet.	Ratio of Vol. of Steam to Vol. of Eq. Weight of Dist. Water at Temp. of Maximum Density.
1	2	3	4	5	6	7	8
p	t	q	L	H	W	V	R
1	102.018	70.040	1043.015	1113.055	.003027	330.4	20623
2	126.302	94.368	1026.094	1120.462	.005818	171.9	10730
3	141.654	109.764	1015.380	1125.144	.008522	117.3	7325
4	153.122	121.271	1007.370	1128.641	.011172	89.51	5588
5	162.370	130.563	1000.899	1131.462	.013781	72.56	4530
6	170.173	138.401	995.441	1133.842	.016357	61.14	3816
7	176.945	145.213	990.695	1135.908	.018908	52.89	3302
8	182.952	151.255	986.485	1137.740	.021436	46.65	2912
9	188.357	156.699	982.690	1139.389	.023944	41.77	2607
10	193.284	161.660	979.232	1140.892	.026437	37.83	2361
11	197.814	166.225	976.050	1142.275	.028911	34.59	2159
12	202.012	170.457	973.098	1143.555	.031376	31.87	1990
13	205.929	174.402	970.346	1144.748	.033828	29.56	1845
14	209.604	178.112	967.757	1145.869	.036265	27.58	1721
14.69	212.000	180.531	966.069	1146.600	.037928	26.37	1646
15	213.067	181.608	965.318	1146.926	.038688	25.85	1614
16	216.347	184.919	963.007	1147.926	.041109	24.33	1519
17	219.452	188.056	960.818	1148.874	.043519	22.98	1434
18	222.424	191.058	958.721	1149.779	.045920	21.78	1359
19	225.255	193.918	956.725	1150.643	.048312	20.70	1292

1	2	3	4	5	6	7	8
p	*t*	*q*	*L*	*H*	*W*	*V*	*R*
20	227.964	196.655	954.814	1151.469	.050696	19.73	1231.0
22	233.069	201.817	951.209	1153.026	.055446	18.04	1126.0
24	237.803	206.610	947.861	1154.471	.060171	16.62	1038.0
26	242.225	211.089	944.730	1155.819	.064870	15.42	962.3
28	246.376	215.293	941.791	1157.084	.069545	14.38	897.6
30	250.293	219.261	939.019	1158.280	.074201	13.48	841.3
32	254.002	223.021	936.389	1159.410	.078839	12.68	791.8
34	257.523	226.594	933.891	1160.485	.083461	11.98	748.0
36	260.883	230.001	931.508	1161.509	.088067	11.36	708.8
38	264.093	233.261	929.227	1162.488	.092657	10.79	673.7
40	267.168	236.386	927.040	1163.426	.097231	10.28	642.0
42	270.122	239.389	924.940	1164.329	.101794	9.826	613.3
44	272.965	242.275	922.919	1165.194	.106345	9.403	587.0
46	275.704	245.061	920.968	1166.029	.110884	9.018	563.0
48	278.348	247.752	919.084	1166.836	.115411	8.665	540.9
50	280.904	250.355	917.260	1167.615	.119927	8.338	520.5
52	283.381	252.875	915.494	1168.369	.124433	8.037	501.7
54	285.781	255.321	913.781	1169.102	.128928	7.756	484.2
56	288.111	257.695	912.118	1169.813	.133414	7.496	467.9
58	290.374	260.002	910.501	1170.503	.137892	7.252	452.7
60	292.575	262.248	908.928	1171.176	.142362	7.024	438.5
62	294.717	264.433	907.396	1171.829	.146824	6.811	425.2
64	296.805	266.566	905.900	1172.466	.151277	6.610	412.6
66	298.842	268.644	904.443	1173.087	.155721	6.422	400.8
68	300.831	270.674	903.020	1173.694	.160157	6.244	389.8
70	302.774	272.657	901.629	1174.286	.164584	6.076	379.3
72	304.669	274.597	900.269	1174.866	.169003	5.917	369.4
74	306.526	276.493	898.938	1175.431	.173417	5.767	360.0
76	308.344	278.350	897.635	1175.985	.177825	5.624	351.1
78	310.123	280.170	896.359	1176.529	.182229	5.488	342.6
80	311.866	281.952	895.108	1177.060	.186627	5.358	334.5
82	313.576	283.701	893.879	1177.580	.191017	5.235	326.8
84	315.250	285.414	892.677	1178.091	.195401	5.118	319.5
86	316.893	287.096	891.496	1178.592	.199781	5.006	312.5
88	318.510	288.750	890.335	1179.085	.204155	4.898	305.8

1	2	3	4	5	6	7	8
p	t	q	L	H	W	V	R
90	320.094	290.373	889.196	1179.569	.208525	4.796	299.4
92	321.653	291.970	888.075	1180.045	.212892	4.697	293.2
94	323.183	293.539	886.972	1180.511	.217253	4.603	287.3
96	324.688	295.083	885.887	1180.970	.221604	4.513	281.7
98	326.169	296.601	884.821	1181.422	.225950	4.426	276.3
100	327.625	298.093	883.773	1181.866	.230293	4.342	271.1
105	331.169	301.731	881.214	1182.945	.241139	4.147	258.9
110	334.582	305.242	878.744	1183.986	.251947	3.969	247.8
115	337.874	308.621	876.371	1184.992	.262732	3.806	237.6
120	341.058	311.885	874.076	1185.961	.273500	3.656	228.3
125	344.136	315.051	871.848	1186.899	.284243	3.518	219.6
130	347.121	318.121	869.688	1187.809	.294961	3.390	211.6
135	350.015	321.105	867.590	1188.695	.305659	3.272	204.2
140	352.827	324.003	865.552	1189.555	.316338	3.161	197.3
145	355.562	326.823	863.567	1190.390	.326998	3.058	190.9
150	358.223	329.566	861.634	1191.200	.337643	2.962	184.9
160	363.346	334.850	857.912	1192.762	.358886	2.786	173.9
170	368.226	339.892	854.359	1194.251	.380071	2.631	164.3
180	372.886	344.708	850.963	1195.671	.401201	2.493	155.6
190	377.352	349.329	847.703	1197.032	.422280	2.368	147.8
200	381.636	353.766	844.573	1198.339	.443310	2.256	140.8
210	385.759	358.041	841.556	1199.597	.464295	2.154	134.5
220	389.736	362.168	838.642	1200.810	.485237	2.061	128.7
230	393.575	366.152	835.828	1201.980	.506139	1.976	123.3
240	397.285	370.008	833.103	1203.111	.527003	1.898	118.5
250	400.883	373.750	830.459	1204.209	.547831	1.825	114.0
260	404.370	377.377	827.896	1205.273	.568626	1.759	109.8
270	407.755	380.905	825.401	1206.306	.589390	1.697	105.9
280	411.048	384.337	822.973	1207.310	.610124	1.639	102.3
290	414.250	387.677	820.609	1208.286	.630829	1.585	99.0
300	417.371	390.933	818.305	1209.238	.651506	1.535	95.8

MISCELLANEOUS TABLES.

STANDARD DIMENSIONS OF WROUGHT-IRON STEAM, GAS, AND WATER PIPES.

Nominal Diameter in Inches.	Thickness in Inches.	Actual Internal Diameter in Inches.	Actual External Diameter in Inches.	Threads per Inch. n	Pitch of Threads.
$\frac{1}{8}$	.068	.270	.405	27	.037
$\frac{1}{4}$	.088	.364	.540	18	.056
$\frac{3}{8}$	.091	.494	.675	18	.056
$\frac{1}{2}$	.109	.623	.840	14	.071
$\frac{3}{4}$	.113	.824	1.050	14	.071
1	.134	1.048	1.315	$11\frac{1}{2}$	.087
$1\frac{1}{4}$	.140	1.380	1.660	$11\frac{1}{2}$	.087
$1\frac{1}{2}$	.145	1.611	1.900	$11\frac{1}{2}$	.087
2	.154	2.067	2.375	$11\frac{1}{2}$	.087
$2\frac{1}{2}$	.204	2.468	2.875	8	.125
3	.217	3.061	3.500	8	.125
$3\frac{1}{2}$	.226	3.548	4.000	8	.125
4	.237	4.026	4.500	8	.125
$4\frac{1}{2}$	.247	4.508	5.000	8	.125
5	.259	5.045	5.563	8	.125
6	.280	6.065	6.625	8	.125
7	.301	7.023	7.625	8	.125
8	.322	7.982	8.625	8	.125
9	.344	9.001	9.688	8	.125
10	.366	10.019	10.750	8	.125

STANDARD PIPE FLANGES.

Inside Diam. of Pipe.	Thickness of Pipe.	Diam. of Bolts.	Length of Bolts.	No. of Bolts.	Thickness of Flange.	Diam. of Bolt Circle.	Diam. of Flange.
2.0	.409	$\frac{5}{8}$	2.0	4	$\frac{5}{8}$	4.75	6.0
2.5	.429	$\frac{5}{8}$	2.25	4	$1\frac{1}{16}$	5.25	7.0
3.0	.448	$\frac{5}{8}$	2.5	4	$\frac{3}{4}$	6.0	7.5
3.5	.466	$\frac{5}{8}$	2.5	4	$\frac{13}{16}$	6.5	8.5
4.0	.486	$\frac{3}{4}$	2.75	4	$\frac{15}{16}$	7.25	9.0
4.5	.498	$\frac{3}{4}$	3.0	8	$\frac{15}{16}$	7.75	9.25
5	.525	$\frac{3}{4}$	3.0	8	$\frac{15}{16}$	8.5	10.0
6	.563	$\frac{3}{4}$	3.0	8	1	9.625	11.0
7	.600	$\frac{3}{4}$	3.25	8	$1\frac{1}{16}$	10.75	12.5
8	.639	$\frac{3}{4}$	3.5	8	$1\frac{1}{8}$	11.75	13.5
9	.678	$\frac{3}{4}$	3.5	12	$1\frac{1}{8}$	13.0	15.0
10	.713	$\frac{7}{8}$	3.625	12	$1\frac{3}{16}$	14.25	16.0
12	.790	$\frac{7}{8}$	3.75	12	$1\frac{1}{4}$	16.5	19.0
14	.864	1	4.25	12	$1\frac{3}{8}$	18.75	21.0
15	.904	1	4.25	16	$1\frac{3}{8}$	20.0	22.25
16	.946	1	4.25	16	$1\frac{7}{16}$	21.25	23.5
18	1.020	$1\frac{1}{8}$	4.75	16	$1\frac{9}{16}$	22.75	25.0
20	1.090	$1\frac{1}{8}$	5.0	20	$1\frac{11}{16}$	25.0	27.5
22	1.180	$1\frac{1}{4}$	5.5	20	$1\frac{13}{16}$	27.25	29.5
24	1.250	$1\frac{1}{4}$	5.5	20	$1\frac{7}{8}$	29.5	32.0
26	1.300	$1\frac{1}{4}$	5.75	24	2	31.75	34.25
28	1.380	$1\frac{1}{4}$	6.0	28	$2\frac{1}{16}$	34.0	36.5
30	1.480	$1\frac{3}{8}$	6.25	28	$2\frac{1}{8}$	36.0	38.75
36	1.710	$1\frac{3}{8}$	6.5	32	$2\frac{3}{8}$	42.75	45.75
42	1.870	$1\frac{1}{2}$	7.25	36	$2\frac{5}{8}$	49.5	52.75
48	2.170	$1\frac{1}{2}$	7.75	44	$2\frac{3}{4}$	56.0	59.5

SPECIFIC HEAT OF SUBSTANCES.

Substance.	Specific Heat.	Substance.	Specific Heat.
Water	1.0000	Ice................	.5040
Sulphur	.2026	Steam (superheated)	.4805
Iron...............	.1138	Air	.2375
Copper	.0951	Oxygen	.2175
Silver	.0570	Hydrogen	3.4090
Tin................	.0562	Carbon monoxide ..	.2479
Mercury	.0333	Carbon dioxide	.2170
Lead	.0314	Nitrogen	.2438

CONSTANTS FOR APPARENT CUT-OFFS USED IN DETERMINING M. E. P.

Cut-off.	Constant.	Cut-off.	Constant.	Cut-off.	Constant.
1/6	.566	3/8	.771	2/3	.917
1/5	.603	.4	.789	.7	.926
1/4	.659	1/2	.847	3/4	.937
.3	.708	.6	.895	.8	.944
1/3	.743	5/8	.904	7/8	.951

RIVETED JOINTS OF BOILERS.

Thickness of Plate. t	Diameter of Rivet.	Diameter of Hole. d	Pitch.		Efficiency of Joint.	
			Single.	Double.	Single.	Double.
1/4″	5/8″	11/16″	2″	3″	.66	.77
5/16″	11/16″	3/4″	2 1/16″	3 1/8″	.64	.76
3/8″	3/4″	13/16″	2 1/8″	3 1/4″	.62	.75
7/16″	13/16″	7/8″	2 3/16″	3 3/8″	.60	.74
1/2″	7/8″	15/16″	2 1/4″	3 1/2″	.58	.73

POSITIONS OF ECCENTRIC RELATIVE TO CRANK.

Kind of Valve.	Kind of Rocker-Arm.	Angle Between Crank and Eccentric.	Position of Eccentric Relative to Crank.
I.... Direct...	Direct.....	90° + angle of advance.....	Ahead of crank.
II... Direct...	Reversing..	90° − angle of advance.....	Behind crank.
III.. Indirect.	Direct.....	90° − angle of advance.....	Behind crank.
IV... Indirect.	Reversing..	90° + angle of advance.....	Ahead of crank.

DIAMETERS OF STEAM AND EXHAUST PIPES.

Diam. of cylinder...	10	12	14	16	18	20	22	24	26	28	30
Diam. of steam pipe.	3	3½	4	4½	5	6	6	7	7	8	9
Diam. of exhaust pipe	3½	4	5	6	6	7	8	9	9	9	10

PISTON SPEEDS OF STEAM ENGINES.

	Ft. per min.
Small stationary engines.......	300 to 600.
Large stationary engines.......	600 to 1,000.
Corliss engines...............	400 to 750.
Locomotives.................	600 to 1,200.

RATIO OF GRATE AREA OF BOILER TO HORSEPOWER.

	Ratio.	Average.
Plain cylindrical.............	.5 to .7	.6
Flue......................	.4 to .5	.45
Multitubular......	.4 to .6	.5
Water tube	.3	.3
Vertical	.6 to .7	.65
Locomotive	.01 to .06	

RATIO OF HEATING SURFACE TO GRATE AREA.

Plain cylindrical...................... 12 to 15
Flue .... 20 to 25
Multitubular .... 25 to 35
Vertical .... 25 to 30
Water tube .... 35 to 40
Locomotive..... .... 50 to 100

RATIO OF HEATING SURFACE TO HORSEPOWER.

Plain cylindrical.................... 6 to 10
Flue.............................. 8 to 12
Multitubular .... 14 to 18
Vertical........................... 15 to 20
Water tube .... 10 to 12
Locomotive....................... 1 to 2

FACTORS OF EVAPORATION.

Temperature of Feedwater.	Gauge Pressures.															
	25	30	35	40	45	50	60	70	80	90	100	120	140	160	180	200
	Factors of Evaporation.															
32	1.204	1.206	1.209	1.211	1.212	1.214	1.217	1.219	1.222	1.224	1.227	1.231	1.234	1.237	1.239	1.241
40	1.196	1.198	1.201	1.203	1.204	1.206	1.209	1.211	1.214	1.216	1.219	1.223	1.226	1.229	1.231	1.233
50	1.185	1.187	1.190	1.192	1.193	1.195	1.198	1.200	1.203	1.205	1.208	1.212	1.215	1.218	1.220	1.222
60	1.175	1.177	1.180	1.182	1.183	1.185	1.188	1.190	1.193	1.195	1.198	1.202	1.205	1.208	1.210	1.212
70	1.165	1.167	1.170	1.172	1.173	1.175	1.178	1.180	1.183	1.185	1.188	1.192	1.195	1.198	1.200	1.202
80	1.154	1.156	1.159	1.161	1.162	1.164	1.167	1.169	1.172	1.174	1.177	1.181	1.184	1.187	1.189	1.191
90	1.144	1.146	1.149	1.151	1.152	1.154	1.157	1.159	1.162	1.164	1.167	1.171	1.174	1.177	1.179	1.181
100	1.134	1.136	1.139	1.141	1.142	1.144	1.147	1.149	1.152	1.154	1.157	1.161	1.164	1.167	1.169	1.171
110	1.123	1.125	1.128	1.130	1.131	1.133	1.136	1.138	1.141	1.143	1.146	1.150	1.153	1.156	1.158	1.160
120	1.113	1.115	1.118	1.120	1.121	1.123	1.126	1.128	1.131	1.133	1.136	1.140	1.143	1.146	1.148	1.150
130	1.102	1.104	1.107	1.109	1.110	1.112	1.115	1.117	1.120	1.122	1.125	1.129	1.132	1.135	1.137	1.139
140	1.092	1.094	1.097	1.099	1.100	1.102	1.105	1.107	1.110	1.112	1.115	1.119	1.122	1.125	1.127	1.129
150	1.082	1.084	1.087	1.089	1.090	1.092	1.095	1.097	1.100	1.102	1.105	1.109	1.112	1.115	1.117	1.119
160	1.071	1.073	1.076	1.078	1.079	1.081	1.084	1.086	1.089	1.091	1.094	1.098	1.101	1.104	1.106	1.108
170	1.061	1.063	1.066	1.068	1.069	1.071	1.074	1.076	1.079	1.081	1.084	1.088	1.091	1.094	1.096	1.098
180	1.050	1.052	1.055	1.057	1.058	1.060	1.063	1.065	1.068	1.070	1.073	1.077	1.080	1.083	1.085	1.087
190	1.040	1.042	1.045	1.047	1.048	1.050	1.053	1.055	1.058	1.060	1.063	1.067	1.070	1.073	1.075	1.077
200	1.030	1.032	1.035	1.037	1.038	1.040	1.043	1.045	1.048	1.050	1.053	1.057	1.060	1.063	1.065	1.067
210	1.020	1.022	1.025	1.027	1.028	1.030	1.033	1.035	1.038	1.040	1.043	1.047	1.050	1.053	1.055	1.057

SIZE OF CHIMNEYS AND HORSEPOWER OF BOILERS.

Diameter in Inches.	Side of Square, Inches.	Actual Area in Sq. Ft.	Effective Area in Sq. Ft.	Height of Chimney in Feet. Commercial Horsepower.										
				50	60	70	80	90	100	110	125	150	175	200
18	16	1.77	0.97	23	25	27								
21	19	2.41	1.47	35	38	41								
24	22	3.14	2.08	49	54	58	62							
27	24	3.98	2.78	65	72	78	83							
30	27	4.91	3.58	84	92	100	107	113						
33	30	5.94	4.47		115	125	133	141	182					
36	32	7.07	5.47		141	152	163	173	219					
39	35	8.30	6.57			183	196	208	258	271				
42	38	9.62	7.76			216	231	245	348	365	389			
48	43	12.57	10.44				311	330	449	472	503	551		
54	48	15.90	13.51				402	427	565	593	632	692	748	
60	54	19.64	16.98				505	539	694	728	776	849	918	981
66	59	23.76	20.83					658	835	876	934	1,023	1,105	1,181
72	64	28.27	25.08					792	995	1,038	1,107	1,212	1,310	1,400
78	70	33.18	29.73						1,163	1,214	1,294	1,418	1,531	1,637
84	75	38.48	34.76						1,344	1,415	1,496	1,639	1,770	1,893
90	80	44.18	40.19						1,537	1,616	1,720	1,876	2,027	2,167
96	86	50.27	46.01											

RULES AND FORMULAS.

FORMULAS USED IN ALGEBRA.

Let a and b be any two quantities, then,

$$(a + b)^2 = a^2 + 2ab + b^2. \qquad \textbf{(1.)} \quad \text{Art. } \textbf{432.}$$

$$(a - b)^2 = a^2 - 2ab + b^2. \qquad \textbf{(2.)} \quad \text{Art. } \textbf{432.}$$

$$(a + b)(a - b) = a^2 - b^2 \qquad \textbf{(3.)} \quad \text{Art. } \textbf{432.}$$

$$a^2 + 2ab + b^2 = (a+b)(a+b) = (a+b)^2. \qquad \textbf{(4.)} \quad \text{Art. } \textbf{455.}$$

$$a^2 - 2ab + b^2 = (a-b)(a-b) = (a-b)^2. \qquad \textbf{(5.)} \quad \text{Ait. } \textbf{455.}$$

$$a^2 - b^2 = (a + b)(a - b). \qquad \textbf{(6.)} \quad \text{Art. } \textbf{462.}$$

Let $ax^2 + bx = c$ be any quadratic equation (it must be borne in mind that b and c may be positive or negative);

then,

$$x = -\frac{b}{2a} \pm \sqrt{\left(\frac{b}{2a}\right)^2 + c} = \frac{-b \pm \sqrt{b^2 + 4ac}}{2a}. \quad \text{Art. } \textbf{597.}$$

THE TRIGONOMETRIC FUNCTIONS.
Art. 754.

Rule 1.—*Sine* $= \dfrac{side\ opposite}{hypotenuse}.$

Rule 2.—*Side opposite* $=$ *hypotenuse* $\times$ *sine.*

Rule 3.—*Cosine* $= \dfrac{side\ adjacent}{hypotenuse}.$

Rule 4.—*Side adjacent* $=$ *hypotenuse* $\times$ *cosine.*

Rule 5.—$Tangent = \dfrac{side\ opposite}{side\ adjacent}$.

Rule 6.—$Side\ opposite = side\ adjacent \times tangent$.

Rule 7.—$Cotangent = \dfrac{side\ adjacent}{side\ opposite}$.

Rule 8.—$Side\ adjacent = side\ opposite \times cotangent$

Rule 9.—$Hypotenuse = \dfrac{side\ opposite}{sine}$.

Rule 10.—$Hypotenuse = \dfrac{side\ adjacent}{cosine}$.

RULES FOR USING TABLE OF LOGARITHMS OF NUMBERS.
Arts. 625–636.

I.   **To find the Characteristic.**—*For a number greater than 1 the characteristic is one less than the number of integral places in the number. For a number wholly decimal the characteristic is negative, and is numerically one greater than the number of ciphers between the decimal point and the first digit of the decimal.*

II.   **To find the Logarithm of a Number not having more than four figures.**—*Find the first three significant figures of the number whose logarithm is desired in the left-hand column; find the fourth figure in the column at the top (or bottom) of the page, and in the column under (or above) this figure, and opposite the first three figures previously found, will be the mantissa, or decimal part, of the logarithm. The characteristic being found as described above, write it at the left of the mantissa, and the resulting expression will be the logarithm of the required number.*

III.   **To find the Logarithm of a Number consisting of five or more figures.**

(**a**)   *If the number consists of more than five figures, and the sixth figure is 5 or greater, increase the fifth figure by 1, and write ciphers in place of the sixth and remaining figures.*

(**b**) *Find the mantissa corresponding to the logarithm of the first four figures, and subtract this mantissa from the next greater mantissa in the table; the remainder is the difference.*

(**c**) *Find in the secondary table headed P. P. a column headed by the same number as that just found for the difference, and in this column opposite the number corresponding to the fifth figure (or fifth figure increased by 1) of the given number (this figure is always situated at the left of the dividing line of the column) will be found the P. P. (proportional part) for that number. The P. P. thus found is to be added to the mantissa found in (**b**), and the result is the mantissa of the logarithm of the given number, as nearly as may be found with five-place tables.*

IV. **To find a Number whose Logarithm is given.** —

(**a**) *Consider the mantissa first. Glance along the different columns of the table which are headed O until the first two figures of the mantissa are found. Then glance down the same column until the third figure is found (or 1 less than the third figure). Having found the first three figures, glance to the right along the row in which they are situated until the last three figures of the mantissa are found. Then the number which heads the column in which the last three figures of the mantissa are found is the fourth figure of the required number, and the first three figures lie in the column headed N, and in the same row in which lie the last three figures of the mantissa.*

(**b**) *If the mantissa cannot be found in the table, find the mantissa which is nearest to, but less than, the given mantissa, and which call the* **next less mantissa.** *Subtract the next less mantissa from the next greater mantissa in the table to obtain the difference. Also subtract the next less mantissa from the mantissa of the given logarithm, and call the remainder the P. P. Looking in the secondary table headed P. P. for the column headed by the difference just found, find the number opposite the P. P. just found (or the P. P. corresponding most* *nearly to that just found) ; this number is the fifth figure of the required number ; the fourth figure will be found at the top of the column containing the next less mantissa, and the first three figures in the column headed N, and in the same row which contains the next less mantissa.*

(c) *Having found the figures of the number as above directed, locate the decimal point by the rules for the characteristic, annexing ciphers to bring the number up to the required number of figures if the characteristic is greater than 4.*

RULES FOR USING TRIGONOMETRIC TABLES.

Given, an angle, to find its sine, cosine, tangent, and cotangent.

Rule 11.—*Find in the table the sine, cosine, tangent, or cotangent corresponding to the degrees and minutes of the angle.*

For the seconds, find the difference of the values of the sine, cosine, tangent, or cotangent taken from the table, between which the seconds of the angle fall ; multiply this difference by a fraction whose numerator is the number of seconds in the given angle, and whose denominator is 60.

If sine or tangent, add this correction to the value first found; if cosine or cotangent, subtract the correction. Art. **756.**

Given, the sine, cosine, tangent, or cotangent to find the angle corresponding.

To find the angle corresponding to a given sine, cosine, tangent, or cotangent whose exact value is not contained in the table:

Rule 12.—*Find the difference of the two numbers in the table between which the given sine, cosine, tangent, or cotangent falls, and use the number of parts in this difference as the denominator of a fraction.*

Find the difference between the number belonging to the **smaller angle,** *and the given sine, cosine, tangent, or cotangent, and use the number of parts in the difference just found as the numerator of the fraction mentioned above. Multiply this fraction by 60, and the result will be the number of seconds to be added to the* **smaller angle.** Art. **758.**

RULES FOR MENSURATION.

THE TRIANGLE.

Rule.—*The area of any triangle equals one-half the product of the base and the altitude.* Art. **766.**

THE QUADRILATERAL.

To find the area of a parallelogram:

Rule.—*The area of any parallelogram equals the product of the base and the altitude.* Art. **777.**

To find the area of a trapezoid:

Rule.—*The area of a trapezoid equals one-half the sum of the parallel sides multiplied by the altitude.* Art. **778.**

To find the area of an irregular figure bounded by straight lines:

Rule.—*Divide the figure into triangles, and find the area of each triangle separately. The sum of the areas of all the triangles will be the area of the figure.* Art. **779.**

THE CIRCLE.

To find the circumference or diameter of a circle:

Rule.—*The circumference of a circle equals the diameter multiplied by 3.1416.* Art. **780.**

Rule.—*The diameter of a circle equals the circumference divided by 3.1416.* Art. **780.**

To find the length of an arc of a circle:

Rule.—*The length of an arc of a circle equals the circumference of the circle of which the arc is a part multiplied by the number of degrees in the arc, and divided by 360.* Art. **781.**

To find the area of a circle:

Rule.—*Square the diameter, and multiply by .7854.* Art. **782.**

Given, the area of a circle to find its diameter:

Rule.—*Divide the area by .7854, and extract the square root of the quotient.* Art. **783.**

To find the area of a sector:

Rule.—*Divide the number of degrees in the arc of a sector by 360. Multiply the result by the area of the circle of which the sector is a part.* Art. **784.**

To find the area of a segment of a circle:

Rule.—*Draw radii from the center of the circle to the extremities of the arc of the segment; find the area of the sector thus formed, subtract from this the area of the triangle formed by the radii and the chord of the arc of the segment, and the result is the area of the segment.* Art. **785.**

THE ELLIPSE.

To find the perimeter of an ellipse:

Rule.—*Multiply the major axis by 1.82, and the minor axis by 1.315. The sum of the results will be the perimeter.* Art. **788.**

To find the exact area of an ellipse:

Rule.—*The area of an ellipse is equal to the product of its two diameters multiplied by .7854.* Art. **789.**

To find the area of any plane figure:

Rule.—*The area of any plane figure may be found by dividing it into triangles, quadrilaterals, circles or parts of circles, and ellipses, finding the area of each part separately, and adding them together.* Art. **790.**

THE PRISM AND CYLINDER.

To find the area of the convex surface of any right prism or right cylinder:

Rule.—*Multiply the perimeter of the base by the altitude.* Art. **803.**

To find the volume of a right prism or cylinder:

Rule.—*The volume of any right prism or cylinder equals the area of the base multiplied by the altitude.* Art. **804.**

THE PYRAMID AND CONE.

To find the area of a right pyramid or right cone:

Rule.—*The convex area of a right pyramid or cone equals the perimeter of the base multiplied by one-half the slant height.* Art. **809.**

To find the volume of a right pyramid or cone:

Rule.—*The volume of a right pyramid or cone equals the area of the base multiplied by one-third of the altitude.* Art. **810.**

THE FRUSTUM OF A PYRAMID OR CONE.

To find the convex area of a frustum of a right pyramid or right cone:

Rule.—*The convex area of a frustum of a right pyramid or right cone equals one-half the sum of the perimeters of its bases multiplied by the slant height of the frustum.* Art. **814.**

To find the volume of the frustum of a pyramid or cone:

Rule.—*Add the areas of the upper base, the lower base, and the square root of the product of the areas of the two bases ; multiply this sum by one-third of the altitude.* Art. **815.**

THE SPHERE.

To find the area of the surface of a sphere:

Rule.—*The area of the surface of a sphere equals the square of the diameter multiplied by 3.1416.* Art. **817.**

To find the volume of a sphere:

Rule.—*The volume of a sphere equals the cube of the diameter multiplied by .5236.* Art. **818.**

FORMULAS USED IN ELEMENTARY MECHANICS.

UNIFORM MOTION.

Let $S =$ the length of space passed over uniformly;

$t =$ the time occupied in passing over the space S;

$V =$ the velocity.

$$V = \frac{S}{t}. \qquad (7.) \quad \text{Art. } \mathbf{859.}$$

$$S = V t. \qquad (8.) \quad \text{Art. } \mathbf{859.}$$

$$t = \frac{S}{V}. \qquad (9.) \quad \text{Art. } \mathbf{859.}$$

MASS, WEIGHT, AND GRAVITY.

If the mass of the body be represented by m, its weight by W, and the force of gravity at the place where the body was weighed by g, we have

$$\text{mass} = \frac{\text{weight of body}}{\text{force of gravity}}, \text{ or } m = \frac{W}{g}. \qquad (10.) \quad \text{Art. } \mathbf{888.}$$

FORMULAS FOR GRAVITY PROBLEMS.

Let W = weight of body at the surface;

w = weight of a body at a given distance above or below the surface;

d = distance between the center of the earth and the center of the body;

R = radius of the earth = 4,000 miles.

Formula for weight when the body is below the surface:

$$w R = d W. \qquad (11.) \quad \text{Art. } \mathbf{891.}$$

Formula for weight when the body is above the surface:

$$w d^2 = W R^2. \qquad (12.) \quad \text{Art. } \mathbf{891.}$$

FALLING BODIES.

Let g = force of gravity = constant accelerating force due to the attraction of the earth;

t = number of seconds the body falls;

v = velocity at the end of the time t;

h = distance that a body falls during the time t.

$$v = g t. \qquad (13.) \quad \text{Art. } \mathbf{896.}$$

That is, the velocity acquired by a freely falling body at the end of t seconds equals 32.16 multiplied by the time in seconds

$$t = \frac{v}{g}. \qquad (14.) \quad \text{Art. } \mathbf{896.}$$

That is, the number of seconds during which a body must have fallen to acquire a given velocity equals the given velocity in feet per second divided by 32.16.

$$h = \frac{v^2}{2g}. \qquad \textbf{(15.)} \quad \text{Art. } \textbf{896.}$$

That is, the height from which a body must fall to acquire a given velocity equals the square of the given velocity divided by 2 ×32.16.

$$v = \sqrt{2gh}. \qquad \textbf{(16.)} \quad \text{Art. } \textbf{896.}$$

That is, the velocity that a body will acquire in falling through a given height equals the square root of the product of twice 32.16 and the given height.

$$h = \tfrac{1}{2}gt^2. \qquad \textbf{(17.)} \quad \text{Art. } \textbf{896.}$$

That is, the distance a body will fall in a given time equals 32.16 ÷ 2 multiplied by the square of the number of seconds.

$$t = \sqrt{\frac{2h}{g}}. \qquad \textbf{(18.)} \quad \text{Art. } \textbf{896.}$$

That is, the time it will take a body to fall through a given height equals the square root of twice the height divided by 32.16.

CENTRIFUGAL FORCE.

The value of the centrifugal force of any revolving body, expressed in pounds, is

$$F = .00034 \, W R N^2; \qquad \textbf{(19.)} \quad \text{Art. } \textbf{903.}$$

in which F = centrifugal force;

$\qquad$ W = total weight of body in pounds;

$\qquad$ R = radius, usually taken as the distance between the center of motion and the center of gravity of the revolving body, in feet;

$\qquad$ N = number of revolutions per *minute*.

THE CENTER OF GRAVITY OF TWO BODIES.

Let l = the distance between the centers of the bodies;

l_1 = the short arm;

w = weight of small body;

W = weight of large body.

$$l_1 = \frac{wl}{W + w}.$$ **(20.)**   Art. **911.**

THE EFFICIENCY OF A MACHINE.

Let F = the force applied to the machine;

V = the velocity ratio of the machine;

W = the weight actually lifted or equivalent resistance overcome;

E = the efficiency of the machine;

Then, $E = \dfrac{W}{FV}.$ **(22.)**   Art. **950.**

WORK.

If the force necessary to overcome the resistance be represented by F, the space through which the resistance acts by S, and the work done by U, then $U = FS$.

If W = the weight of a body, and h = the height through which it is raised, $U = Wh$. Hence the work done

$$U = FS = Wh.$$ **(23.)**   Art. **953.**

POWER.

The power of a machine may always be determined by *dividing the work done in foot-pounds by the time in minutes required to do the work; i. e.*,

$$\text{Power} = \frac{FS}{T}.$$ **(24.)**   Art. **954.**

KINETIC ENERGY.

Let W = the weight of the body in pounds;

v = its velocity in feet per second;

h = the height in feet through which the body must fall to produce the velocity v;

m = the mass of the body = $\dfrac{W}{g}$. (See formula **10.**)

The work necessary to raise a body through a height h is Wh. The velocity produced in falling a height h is

$$v = \sqrt{2\,g\,h}, \text{ and } h = \frac{v^2}{2\,g}. \quad \text{(See formulas 15 and 16.)}$$

Therefore, work $= Wh = W\dfrac{v^2}{2\,g} = \frac{1}{2} \times \dfrac{W}{g} \times v^2 = \frac{1}{2}\,m\,v^2$,

or $\qquad Wh = \frac{1}{2}\,m\,v^2.$ **(25.)** Art. **957.**

DENSITY.

The **density of a body** is its mass divided by its volume in cubic feet.

Let D be the density; then the density of a body is $D = \dfrac{m}{V}.$ Since $m = \dfrac{W}{g}$, $D = \dfrac{W}{g\,V}.$ **(26.)** Art. **962.**

RULES AND FORMULAS USED IN HYDRAULICS.

PASCAL'S LAW.

Rule.—*The pressure per unit of area exerted anywhere upon a mass of liquid is transmitted undiminished in all directions, and acts with the same force upon all surfaces in a direction at* **right angles** *to those surfaces.* Art. **968.**

THE GENERAL LAW FOR THE DOWNWARD PRESSURE UPON THE BOTTOM OF ANY VESSEL.

Rule.—*The pressure upon the bottom of a vessel containing a fluid is independent of the shape of the vessel, and is equal to the weight of a prism of the fluid whose base has the same area as the bottom of the vessel, and whose altitude is the distance between the bottom and the upper surface of the fluid plus the pressure per unit of area upon the upper surface of the fluid, multiplied by the area of the bottom of the vessel.* Art. **971.**

GENERAL LAW FOR UPWARD PRESSURE.

Rule.—*The upward pressure on any submerged horizontal surface equals the weight of a prism of the liquid whose base has an area equal to the area of the submerged surface,* *and whose altitude is the distance between the submerged surface and the upper surface of the liquid plus the pressure per unit of area on the upper surface of the fluid, multiplied by the area of the submerged surface.* Art. **973.**

GENERAL LAW FOR LATERAL PRESSURE.

Rule.—*The pressure upon any vertical surface due to the weight of a liquid is equal to the weight of a prism of the liquid whose base has the same area as the vertical surface, and whose altitude is the depth of the center of gravity of the vertical surface below the level of the liquid.*

Any additional pressure is to be added, as in the previous cases. Art. **975.**

GENERAL LAW FOR PRESSURE.

Rule.—*The pressure exerted by a fluid in any direction upon* **any** *surface is equal to the weight of a prism of the fluid whose base is the projection of the surface at right angles to the direction considered, and whose height is the depth of the center of gravity of the surface below the level of the liquid.* Art. **979.**

SPECIFIC GRAVITY.

Let W be the weight of the solid in air and W' the weight in water; then, $W - W' =$ the weight of a volume of water equal to the volume of the solid, and

$$\text{Sp. Gr.} = \frac{W}{W - W'}. \qquad (27.) \quad \text{Art. } \mathbf{982.}$$

If the body be lighter than water, a piece of iron or other heavy substance must be attached to it sufficiently heavy to sink both. *Then weigh both bodies in air and both in water.*

Let W = weight of both bodies in air;
$\quad W'$ = weight of both bodies in water;
$\quad w$ = weight of light body in air;
$\quad W_1$ = weight of heavy body in air;
$\quad W_2$ = weight of heavy body in water.

Then, the specific gravity of the light body is given by

$$\text{Sp. Gr.} = \frac{w}{(W - W') - (W_1 - W_2)}. \qquad (27a.) \quad \text{Art. } \mathbf{983.}$$

To find the specific gravity of a liquid:

Weigh an empty flask; fill it with water, then weigh it, and find the difference between the two results; this will equal the weight of the water. Then weigh the flask filled with the liquid, and subtract the weight of the flask; the result is the weight of a volume of the liquid equal to the volume of the water. The weight of the liquid divided by the weight of the water is the specific gravity of the liquid.

Let W = the weight of the flask and liquid;
W'' = the weight of the flask and water;
w = the weight of the flask.

Then, $\text{Sp. Gr.} = \dfrac{W - w}{W'' - w}.$ $(27b.)$ Art. **984.**

FORMULAS FOR FLOW OF WATER.

MEAN VELOCITY.

Let Q = the quantity in cubic feet which passes any section in 1 second;
A = the area of the section in square feet;
v_m = the mean velocity in feet per second.

Then, $Q = A v_m,$ $(28a.)$ Art. **989.**

and $v_m = \dfrac{Q}{A}.$ $(28b.)$ Art. **989.**

VELOCITY OF EFFLUX FROM AN ORIFICE.

Let v = the velocity of efflux in feet per second;
h = the head in feet on the orifice considered;
h_1 = the head equivalent to a pressure p;
W = the weight of the water in pounds flowing through the aperture per second.

The kinetic energy of the issuing water $= \dfrac{W v^2}{2 g}.$

The work the issuing water can do $= Wh$.

$$Wh = \frac{Wv^2}{2g}, \text{ or } v = \sqrt{2gh}.$$

$h_1 = \frac{p}{.434}$, where h_1 is in feet, and p in pounds per square inch.

$h_1 = \frac{p}{62.5}$, where h_1 is in feet, and p in pounds per square foot.

$h + h_1 =$ the *total head*.

$$v = \sqrt{2g(h_1 + h)}. \qquad (29.) \quad \text{Art. } \mathbf{991.}$$

If a is the area of a large orifice in the bottom of a small vessel whose area is A, the velocity is

$$v \sqrt{\frac{2gh}{1 - \frac{a^2}{A^2}}}. \qquad (31.) \quad \text{Art. } \mathbf{993.}$$

THEORETICAL RANGE OF A JET.

Let $h =$ head on center of orifice;

$y =$ vertical height of orifice above the surface where the water strikes;

$R =$ range.

Then, $R = \sqrt{4hy}. \qquad (30.) \quad \text{Art. } \mathbf{992.}$

FLOW THROUGH ORIFICES.

Velocity of the Jet.

Let $v =$ theoretical velocity;

$v' =$ actual maximum velocity;

$c' =$ coefficient of velocity;

$h =$ head on center of orifice;

$g =$ acceleration due to gravity $= 32.16$.

$$v' = c'v = c'\sqrt{2gh}. \qquad (32.) \quad \text{Art. } \mathbf{994.}$$

An average value of c' is .98.

Discharge of an Orifice.

Let Q = theoretical discharge;
Q' = actual discharge;
a = area of orifice;
c'' = coefficient of discharge;
h = head on center of orifice;
g = acceleration due to gravity = 32.16.

An average value of c'' is .61. Then,

$$Q' = c'' Q = c'' a \sqrt{2gh} = .61\, a \sqrt{2gh}. \qquad (33.) \quad \text{Art. } 994.$$

Discharge of Standard Orifices.

Let Q = discharge in cubic feet per second;
d = diameter of a circular or length of a side of a square orifice in feet;
d' = depth of a rectangular orifice in feet;
b = breadth of a rectangular orifice in feet;
h = head on the center of a circular or of a square orifice in feet;
h_1 = head on the upper edge of a rectangular orifice in feet;
h_2 = head on the lower edge of a rectangular orifice in feet;
c = coefficient of discharge (see tables of Coefficients of Discharge for Standard Orifices);
g = acceleration due to gravity = 32.16.

For a circular vertical orifice,

$$Q = .7854\, d^2 c \sqrt{2gh} = 6.299\, d^2 c \sqrt{h}. \qquad (34a.) \quad \text{Art. } 996.$$

For a square vertical orifice,

$$Q = c\, d^2 \sqrt{2gh} = 8.02\, c\, d^2 \sqrt{h}. \qquad (34b.) \quad \text{Art. } 997.$$

For a rectangular vertical orifice,

$$Q = c \times \tfrac{2}{3} b \sqrt{2g}\,(\sqrt{h_2^3} - \sqrt{h_1^3}) =$$
$$5.347\, c\, b\,(\sqrt{h_2^3} - \sqrt{h_1^3}). \qquad (34c.) \quad \text{Art. } 998.$$

If the head h on the center of a rectangular vertical

orifice is greater than $4\,d$, the discharge may be computed by the formula

$$Q = c\,b\,d\sqrt{2gh} = 8.02\,c\,b\,d\sqrt{h}. \qquad (34d.) \quad \text{Art. } \mathbf{998.}$$

For approximate computations the value of c that may be used in formulas $34c$ and $34d$ is $c = .615$.

Discharge of a Submerged Rectangular Orifice.

Let Q = discharge in cubic feet per second;

b = breadth of orifice in feet;

d = depth of orifice in feet;

h_o = the difference in the level of the water on the two sides of the orifice in feet;

g = acceleration due to gravity = 32.16.

Then,

$$Q = .615\,b\,d\sqrt{2gh_o} = 4.932\,b\,d\sqrt{h_o}. \qquad (34e.) \quad \text{Art. } \mathbf{999.}$$

THE DISCHARGE OF WEIRS.

Let l = length of the weir in feet;

H = measured head in feet;

v = velocity with which the water approaches the weir in feet per second;

h = head equivalent to the velocity with which the water approaches the weir in feet;

c = coefficient of discharge (see tables of Coefficients of Discharge for Weirs);

Q = theoretical discharge in cubic feet per second;

Q' = actual discharge in cubic feet per second.

The theoretical discharge per second is

$$Q = \tfrac{2}{3}\sqrt{2g}\,l(H+h)^{\frac{3}{2}}. \qquad (35a.) \quad \text{Art. } \mathbf{1006.}$$

If there is no velocity of approach, this becomes

$$Q = \tfrac{2}{3}\sqrt{2g}\,l\,H^{\frac{3}{2}}. \qquad (35b.) \quad \text{Art. } \mathbf{1006.}$$

The actual discharge for weirs with end contractions is given by the formulas

$$Q' = c\left(\tfrac{2}{3}\sqrt{2g}\right)l(H+\tfrac{1}{2}h)^{\frac{3}{2}} = 5.347\,c\,l(H+\tfrac{1}{2}h)^{\frac{3}{2}}, \qquad (36a.) \quad \text{Art. } \mathbf{1006.}$$

and

$$Q' = c\left(\tfrac{2}{3}\sqrt{2g}\right)l\,H^{\frac{3}{2}} = 5.347\,c\,l\,H^{\frac{3}{2}}. \qquad (36b.) \quad \text{Art. } \mathbf{1006.}$$

For weirs without end contractions, the formulas are

$$Q' = c\tfrac{2}{3}\sqrt{2g}\,l\,(H + 1.4\,h)^{\frac{3}{2}} = 5.347\,c\,l\,(H + 1.4\,h)^{\frac{3}{2}},$$
$$\text{(37a.)} \quad \text{Art. } \mathbf{1006.}$$

and $\quad Q' = c\tfrac{2}{3}\sqrt{2g}\,l\,H^{\frac{3}{2}} = 5.347\,c\,l\,H^{\frac{3}{2}}.$ (37b.) Art. **1006.**

The **velocity of approach** is the mean velocity with which the water flows through the canal leading to the weir. If A is the area of the cross-section of the water in this canal, we have $v = \dfrac{Q'}{A}$, from which we see that Q' must be determined approximately by assuming $v = 0$, and then use this value of Q' to find v. V may also be measured approximately by means of a float on the water in the canal or stream.

Having found v, we have the equivalent head $h = \dfrac{v^2}{2g} = .01555\,v^2$. (See Arts. **990** and **991**.) Since v is small with a properly constructed weir, it is usually neglected, unless great accuracy is required.

FLOW OF WATER THROUGH PIPES.

Let l = length of pipe in feet;

$\quad d$ = diameter of pipe in feet;

$\quad d_1$ = diameter of pipe in inches;

$\quad v$ = mean velocity of flow through pipe in feet per second;

$\quad h$ = total head on outlet end of pipe in feet;

$\quad h''$ = head in feet equivalent to the velocity v;

$\quad h'''$ = head in feet equivalent to the loss of pressure at entrance to pipe;

$\quad h^{\mathrm{IV}}$ = head in feet equivalent to the loss in pressure produced by friction in pipe;

$\quad h^{\mathrm{V}}$ = head in feet equivalent to loss in pressure produced by angular bends in pipe;

$\quad h^{\mathrm{VI}}$ = head in feet equivalent to loss in pressure produced by circular bends in pipe;

$\quad f$ = a coefficient for loss of head due to friction (see table of Coefficients f for Smooth Iron Pipes);

$\quad m$ = a coefficient for loss of head at entrance;

$\quad n$ = number of bends in pipe;

c = a coefficient for loss of head due to angular bends (see table of Coefficients for Angular Bends);

c' = a coefficient for loss of head due to circular bends (see table of Coefficients for Circular Bends);

Q = quantity discharged by pipe in cubic feet per second;

Q' = quantity discharged by pipe in gallons per second;

r = radius of pipe in feet;

R = radius of circular bend in pipe in feet;

$a°$ = number of degrees of angular bend in pipe.

General Formulas.

Loss of head at entrance,

$$h''' = m\,h'' = m\frac{v^2}{2g}. \qquad (39.) \quad \text{Art. } \mathbf{1020.}$$

Loss of head due to friction,

$$h^{IV} = f\frac{l}{d}\frac{v^2}{2g}. \qquad (40a.) \quad \text{Art. } \mathbf{1021.}$$

Loss of head due to angular bends,

$$h^{V} = c\frac{v^2}{2g}. \qquad (40b.) \quad \text{Art. } \mathbf{1023.}$$

Loss of head due to circular bends,

$$h^{VI} = c'\frac{v^2}{2g}. \qquad (40c.) \quad \text{Art. } \mathbf{1023.}$$

Total head,

$$h = h'' + h''' + h^{IV} + h^{V} =$$
$$\frac{v^2}{2g} + f\frac{l}{d}\frac{v^2}{2g} + m\frac{v^2}{2g} + n\,c'\frac{v^2}{2g}. \qquad (41a.) \quad \text{Art. } \mathbf{1024.}$$

Velocity of flow,

$$v = \sqrt{\frac{2gh}{1 + f\frac{l}{d} + m + nc'}} =$$
$$8.02 \sqrt{\frac{h}{1 + f\frac{l}{d} + m + nc'}}. \qquad (42.) \quad \text{Art. } \mathbf{1024.}$$

If $m = .5$ and there are no sharp bends,

$$v = \sqrt{\frac{2gh}{1.5 + f\frac{l}{d}}} = 8.02 \sqrt{\frac{h}{1.5 + f\frac{l}{d}}}; \quad \text{(43.)} \quad \text{Art. } \mathbf{1024.}$$

and, when the diameter is in inches,

$$v = 2.315 \sqrt{\frac{h\, d_1}{fl + .125\, d_1}}. \quad \text{(43a.)} \quad \text{Art. } \mathbf{1025.}$$

Velocity Through Long Pipes.

When the diameter is in feet,

$$v = 8.02 \sqrt{\frac{d\, h}{fl}}. \quad \text{(44.)} \quad \text{Art. } \mathbf{1025.}$$

When the diameter is in inches,

$$v = 2.315 \sqrt{\frac{h\, d_1}{fl}}. \quad \text{(44a.)} \quad \text{Art. } \mathbf{1025.}$$

Head Required to Produce a Given Velocity.

General formula,

$$h = \frac{v^2 \left(1 + f\frac{l}{d} + m + n\, c'\right)}{64.32}. \quad \text{(45.)} \quad \text{Art. } \mathbf{1026.}$$

When the influence of bends is neglected and m has the value .5, the formula is

$$h = \frac{f\, l\, v^2}{64.32\, d} + .0233\, v^2. \quad \text{(45a.)} \quad \text{Art. } \mathbf{1026.}$$

When the diameter is given in inches,

$$h = \frac{f\, l\, v^2}{5.36\, d} + .0233\, v^2. \quad \text{(45b.)} \quad \text{Art. } \mathbf{1026.}$$

The Quantity Discharged from Pipes.

When the diameter is given in feet, the discharge in cubic feet per second is

$$Q = .7854\, d^2\, v. \quad \text{(46.)} \quad \text{Art. } \mathbf{1027.}$$

Since one cubic foot contains 7.48 gallons, if the diameter is in feet, we have

$Q' = .7854 \, d^2 \, v \times 7.48$ gallons per second;   (46a.)

Art. **1027.**

and for the diameter in inches,

$Q' = .0408 \, d_1^2 \, v$ gallons per second.   (46b.)   Art. **1027.**

The Diameter of Pipes.

With h, l, and d in feet and the quantity Q in cubic feet per second, the formula for the diameter of a pipe without sharp bends is

$$d = 0.479 \left[(1.5 \, d + f l) \frac{Q^2}{h} \right]^{\frac{1}{5}}.$$ (47.)   Art. **1028.**

In using this formula, take the approximate value of f as .02, and compute an approximate value for d, neglecting the term $1.5 \, d$ in the second member of the formula. With this value of d, find the value of v from the formula $v = \dfrac{Q}{.7854 \, d^2}$, and find the corresponding value of f from the table of Coefficients for Pipes.

Repeat the computation for d by placing the approximate values of d and f just found in the second member of the formula. One or two repetitions of this process will give a near approximation of d from which to select the pipe from the standard market sizes.

For pipes whose length is more than 4,000 times their diameter, the following formula may be used:

$$d = 0.479 \left(\frac{f l Q^2}{h} \right)^{\frac{1}{5}}.$$ (47a.)   Art. **1028.**

FLOW OF WATER IN CONDUITS AND CHANNELS.

Let S = slope of a conduit or channel;

h = a given fall;

l = distance in which the fall h occurs;

p = wetted perimeter;

a = area of water cross-section;

r = hydraulic radius;

v = mean velocity of flow;

Q = quantity discharged;

$c =$ a coefficient to be determined by Kutter's formula;

$n =$ coefficient of roughness to be used in Kutter's formula (see table of Coefficients of Roughness).

Formula for slope,

$$S = \frac{h}{l}. \qquad \text{(48.)} \quad \text{Art. } \mathbf{1032.}$$

Hydraulic radius,

$$r = \frac{a}{p}. \qquad \text{(49.)} \quad \text{Art. } \mathbf{1032.}$$

Discharge,

$$Q = a\,v. \qquad \text{Art. } \mathbf{1032.}$$

Mean velocity,

$$v = c \sqrt{r\,S}. \qquad \text{(50.)} \quad \text{Art. } \mathbf{1033.}$$

To find the value of c use **Kutter's formula,**

$$c = \frac{23 + \dfrac{1}{n} + \dfrac{.00155}{S}}{.5521 + \left(23 + \dfrac{.00155}{S}\right)\dfrac{n}{\sqrt{r}}}. \qquad \text{(51.)} \quad \text{Art. } \mathbf{1033.}$$

The value of n to be used in this formula is to be taken from the following table to correspond with the character of the channel:

VALUES OF THE COEFFICIENT OF ROUGHNESS.

For Use in Kutter's Formula.

Character of Channel.	Value of n.
Clean, well-planed timber	.009
Clean, smooth, glazed iron and stoneware pipes	.010
Masonry smoothly plastered with cement	.011
Clean, smooth cast-iron pipe	.011
Ordinary cast-iron pipe	.012
Unplaned timber	.012
Selected sewer pipes, well laid and thoroughly flushed.	.012
Rough iron pipes.	.013
Ordinary sewer pipes laid under usual conditions	.013
Dressed masonry and well-laid brickwork	.015

| Character of Channel. | Value of n. |

Good rubble masonry and ordinary rough or fouled
brickwork 017

Coarse rubble masonry 020

Gravel, compact and firm 020

Earth canals, well made and in good alinement 0225

Rivers and canals in moderately good order and per-
fectly free from stones and weeds. 025

Rivers and canals in rather bad condition and some-
what obstructed by stones and weeds. 030

Rivers and canals in bad condition, overgrown with
vegetation and strewn with stones and other
detritus, according to condition . . . . .035 to .050

FORMULAS USED IN PNEUMATICS.

**PRESSURE, VOLUME, DENSITY, AND WEIGHT OF AIR
WHEN THE TEMPERATURE IS CONSTANT:**

Marlotte's Law.—*The temperature remaining the same,
the volume of a given quantity of gas varies inversely as the
pressure.*

Let p = pressure for one position of the piston;

p_1 = pressure for any other position of the piston;

v = volume corresponding to the pressure p;

v_1 = volume corresponding to the pressure p_1.

Then, $p\,v = p_1\,v_1$.   **(53.)**   Art. **1049.**

Let D be the density corresponding to the pressure p and
volume v, and D_1 be the density corresponding to the
pressure p_1 and volume v_1; then,

$p : D = p_1 : D_1$, or $p\,D_1 = p_1\,D$,   **(54.)**   Art. **1052.**

and $v : D_1 = v_1 : D$, or $v\,D = v_1\,D_1$.   **(55.)**   Art. **1052.**

Thus, let W be the weight of a cubic foot of air or other gas,
whose volume is v, and pressure is p; let W_1 be the weight
of a cubic foot when the volume is v_1, and pressure is p_1;
then,

$p\,W_1 = p_1\,W$.   **(56.)**   Art. **1052.**

$v\,W = v_1\,W_1$.   **(57.)**   Art. **1052.**

Gay-Lussac's Law.—*If the pressure remains constant, every increase of temperature of 1° F. produces in a given quantity of gas an expansion of $\frac{1}{492}$ of its volume at 32° F.*

If the pressure remains constant it will also be found that every decrease of temperature of 1° F. will cause a decrease of $\frac{1}{492}$ of the volume at 32° F.

Let v = original volume of gas;

v_1 = final volume of gas;

t = temperature corresponding to volume v;

t_1 = temperature corresponding to volume v_1.

Then, $$v_1 = v \left(\frac{460 + t_1}{460 + t} \right). \quad \textbf{(58.)} \quad \text{Art. } \textbf{1054.}$$

That is, *the volume of gas after heating (or cooling) equals the original volume multiplied by 460 plus the final temperature divided by 460 plus the original temperature.*

Let p = the original tension;

t = the corresponding temperature;

p_1 = final tension;

t_1 = final temperature.

Then, $$p_1 = p \left(\frac{460 + t_1}{460 + t} \right). \quad \textbf{(59.)} \quad \text{Art. } \textbf{1055.}$$

Let p = pressure in pounds per square inch;

V = volume of air in cubic feet;

T = absolute temperature;

W = weight in pounds.

Then, $$p V = .37052 \, T. \quad \textbf{(60.)} \quad \text{Art. } \textbf{1056.}$$

If the weight of the air be greater or less than 1 pound, the following formula must be used:

$$p V = .37052 \, W T. \quad \textbf{(61.)} \quad \text{Art. } \textbf{1057.}$$

Let p_1, V_1, and T_1 represent the pressure, volume, and temperature of the same weight of air in another state; then,

$$\frac{p V}{T} = \frac{p_1 V_1}{T_1}. \quad \textbf{(62.)} \quad \text{Art. } \textbf{1058.}$$

MIXTURE OF TWO GASES HAVING UNEQUAL VOLUMES AND PRESSURES.

Let v and p be the volume and pressure, respectively, of one of the gases.

Let v_1 and p_1 be the volume and pressure, respectively, of the other gas.

Let V and P be the volume and pressure, respectively, of the mixture. Then, if the temperature remains the same,

$$VP = vp + v_1 p_1. \qquad \textbf{(63.)} \quad \text{Art. } \textbf{1062.}$$

MIXTURE OF TWO VOLUMES OF AIR HAVING UNEQUAL PRESSURES, VOLUMES, AND TEMPERATURES.

If a body of air having a temperature t_1, a pressure p_1, and a volume v_1 be mixed with another volume of air having a temperature t_2, a pressure p_2, and a volume v_2, to form a volume V having a pressure P, and a temperature t, then, either the new temperature t, the new volume V, or the new pressure P may be found, if the other two quantities are known, by the following formula, in which T_1, T_2, and T are the absolute temperatures corresponding to t_1, t_2, and t

$$PV = \left[\frac{p_1 v_1}{T_1} + \frac{p_2 v_2}{T_2} \right] T. \qquad \textbf{(64.)} \quad \text{Art. } \textbf{1063.}$$

FORMULAS USED IN STRENGTH OF MATERIALS.

UNIT STRESS, UNIT STRAIN, AND COEFFICIENT OF ELASTICITY.

Let P = the total stress in pounds;

A = area of cross-section in square inches;

S = unit stress in pounds per square inch;

l = length of body in inches;

e = elongation in inches;

s = unit strain;

E = coefficient of elasticity.

$$S = \frac{P}{A}, \text{ or } P = A S. \qquad \textbf{(65.)} \quad \text{Art. } \textbf{1103.}$$

$$s = \frac{e}{l}, \text{ or } e = l\,s. \qquad (66.) \quad \text{Art. } \mathbf{1104.}$$

$$E = \frac{S}{s} = \frac{P}{A} \div \frac{e}{l} = \frac{Pl}{A\,e}. \qquad (67.) \quad \text{Art. } \mathbf{1110.}$$

STRENGTH OF PIPES AND CYLINDERS.

Let d = inside diameter of pipe in inches;

l = length of pipe in inches;

p = pressure in pounds per square inch;

P = total pressure; then, $P = p\,l\,d$;

t = thickness of pipe;

S = working strength of the material.

For longitudinal rupture

$p\,l\,d = 2\,t\,l\,S$, or

$p\,d = 2\,t\,S.$ (68.)   Art. **1123.**

For transverse rupture

$p\,d = 4\,t\,S.$ (69.)   Art. **1124.**

Since, for longitudinal rupture, $p\,d = 2\,t\,S$, it is seen that a cylinder is twice as strong against transverse rupture as against longitudinal rupture.

For pipes and cylinders whose thickness is greater than $\frac{1}{10}$ of the radius, use the following formula, in which r = the inner radius, and the other letters have the same meaning as before.

$$p = \frac{S\,t}{r + t}. \qquad (70.) \quad \text{Art. } \mathbf{1125.}$$

The following formula gives the collapsing pressure in lb per sq. in. for wrought-iron pipe:

$$p = 9,600,000 \frac{t^{3.18}}{l\,d}. \qquad (71.) \quad \text{Art. } \mathbf{1126.}$$

MOMENT OF INERTIA, RESISTING MOMENT, AND BENDING MOMENT OF BEAMS.

Let I = moment of inertia;

A = area of cross-section;

r = radius of gyration;

c = distance from neutral axis to outermost fiber;

S_1 = ultimate strength of flexure;

f = factor of safety;

M = bending moment.

$$I = A\,r^2. \qquad (72.) \quad \text{Art. } 1154.$$

The resisting moment is given by the expression

$$\frac{S}{c}\,A\,r^2 = \frac{S}{c}\,I, \text{ or } S\frac{I}{c}.$$

For the bending moment

$$M = S_1\frac{I}{c}, \qquad (73.) \quad \text{Art. } 1156.$$

and, when a factor of safety is used,

$$M = \frac{S_1 I}{f\,c}. \qquad (74.) \quad \text{Art. } 1159.$$

DEFLECTION OF A BEAM.

Let a = a constant depending on the manner of loading the beam and the condition of the ends;

s = the deflection;

E = the coefficient of elasticity;

l = the length of the beam in inches;

W = the total weight supported in pounds;

I = moment of inertia about the neutral axis.

$$s = a\frac{W\,l^3}{E\,I}. \qquad (75.) \quad \text{Art. } 1162.$$

STRENGTH OF COLUMNS.

Let W = load on a column;

S_2 = ultimate strength for compression;

A = area of section of column;

f = factor of safety;

l = length of column in inches;

g = a constant to be taken from table;

I = least moment of inertia of cross-section;

b = length of longer side of a rectangular column;

d = length of shorter side of a rectangular column, or the diameter of a circular column;

c = length of one side of a square column.

$$W = \frac{S_1 A}{f\left(1 + \frac{A l^2}{g I}\right)}. \qquad \textbf{(76.)} \quad \text{Art. } \mathbf{1169.}$$

$$c = \sqrt{\frac{W f}{2 S_1} + \sqrt{\frac{W f}{S_1}\left(\frac{W f}{4 S_1} + \frac{12 l^2}{g}\right)}}. \qquad \textbf{(77.)} \quad \text{Art. } \mathbf{1171.}$$

For a circular column,

$$d = 1.4142 \sqrt{\frac{.3183 \, W f}{S_1} + \sqrt{\frac{.3183 \, W f}{S_1}\left(\frac{.3183 \, W f}{S_1} + \frac{16 \, l^2}{g}\right)}}.$$
$$\textbf{(78.)} \quad \text{Art. } \mathbf{1171.}$$

For a rectangular column, assume d, then,

$$b = \frac{W f\left(1 + \frac{12 \, l^2}{d^2 g}\right)}{d S_1}. \qquad \textbf{(79.)} \quad \text{Art. } \mathbf{1171.}$$

STRENGTH OF SHAFTS.

Let d = diameter of a round shaft, or side of a square shaft, in inches;

c = a constant (see table of Constants for Shafting);

c_1 = a constant (see table of Constants for Shafting);

P = a force applied at the end of a lever arm in pounds;

r = length of lever arm in inches;

H = horsepower transmitted by shaft;

N = number of revolutions per minute;

k = a constant (see table of Constants for Shafting);

k_1 = a constant (see table of Constants for Shafting);

q = a constant (see table of Constants for Shafting);

q_1 = a constant (see table of Constants for Shafting).

For all solid shafts below 11 inches in diameter use the formula

$$d = c \sqrt[3]{P r} = c_1 \sqrt[3]{\frac{H}{N}}. \qquad \textbf{(80.)} \quad \text{Art. } \mathbf{1173.}$$

If the diameter of a wrought-iron shaft is greater than 12.4″, of a cast-iron shaft greater than 10.3″, or of a steel shaft greater than 13.6″, use the following formula:

$$d = k \sqrt[3]{Pr} = k_1 \sqrt[3]{\frac{H}{N}}. \qquad (81.) \quad \text{Art. } 1174.$$

For a hollow (round) shaft use formula **82** or **83**.

$$P = q \left(\frac{d_1^4 - d_2^4}{d_1 r} \right), \qquad (82.) \quad \text{Art. } 1174.$$

or $\quad H = q_1 N \left(\dfrac{d_1^4 - d_2^4}{d_1} \right). \qquad (83.) \quad \text{Art. } 1174.$

CONSTANTS FOR SHAFTING.

VALUES OF c AND c_1 TO BE USED IN FORMULA 80.

Material.	c		c_1	
	Round.	Square.	Round.	Square.
Wrought Iron.	.310	.272	4.92	4.31
Cast Iron.	.353	.309	5.59	4.89
Steel.	.297	.260	4.70	4.11

VALUES OF k, k_1, q, AND q_1 TO BE USED IN FORMULAS 81, 82, AND 83.

Material.	k	k_1	q	q_1
Wrought Iron.	.0909	3.62	1,335	.0212
Cast Iron.	.1145	4.56	669	.0106
Steel.	.0828	3.30	1,767	.0280

STRENGTH OF ROPES AND CHAINS.

Let $P =$ working or safe load in pounds;

$\quad C =$ circumference of rope in inches;

$\quad d =$ diameter of the link of a chain in inches.

For manila ropes, hemp ropes, or tarred hemp ropes,

$$P = 100\, C^2. \qquad \textbf{(84.)} \quad \text{Art. } \textbf{1175.}$$

For iron wire rope of 7 strands, 19 wires to the strand,

$$P = 600\, C^2. \qquad \textbf{(85.)} \quad \text{Art. } \textbf{1176.}$$

For the best steel wire rope, 7 strands, 19 wires to the strand,

$$P = 1,000\, C^2. \qquad \textbf{(86.)} \quad \text{Art. } \textbf{1176.}$$

For open-link chains made from a good quality of wrought iron,

$$P = 12,000\, d^2, \qquad \textbf{(87.)} \quad \text{Art. } \textbf{1179.}$$

and for stud-link chains,

$$P = 18,000\, d^2. \qquad \textbf{(88.)} \quad \text{Art. } \textbf{1179.}$$

FORMULAS USED IN SURVEYING.

RADIUS OF A CURVE.

To find the radius, the degree being given:

Let R = the length of the required radius;

$\quad D$ = the deflection angle equal to one-half the degree of the given curve.

$$R = \frac{50}{\sin D}. \qquad \textbf{(89.)} \quad \text{Art. } \textbf{1249.}$$

LENGTH OF SUB-CHORDS.

For curves of short radii:

Let C = the length of the required chord;

$\quad R$ = the radius of the given curve;

$\quad D$ = the deflection angle of the given curve, equal to one-half its degree.

$$C = 2\,R \sin D. \qquad \textbf{(90.)} \quad \text{Art. } \textbf{1250.}$$

LENGTH OF THE TANGENT OF A CURVE.

When the radius and intersection angle are given:

Let T = the length of the required tangent;

$\quad R$ = the radius of the given curve;

$\quad I$ = the intersection angle of the given curve.

$$T = R \tan \tfrac{1}{2}\, I. \qquad \textbf{(91.)} \quad \text{Art. } \textbf{1251.}$$

CHORD DEFLECTION.

When the length of the chord and the radius are given:

Let d = the required chord deflection;

    c = the length of the chord of the given curve;

    R = the radius of the given curve.

$$d = \frac{c^2}{R}. \qquad \textbf{(92.)} \quad \text{Art. } \textbf{1255.}$$

TANGENT DEFLECTION.

When the length of the tangent, or of its corresponding chord, and the radius are given:

Let c = the length of the tangent or corresponding chord;

    R = the radius of the given curve.

$$\text{tangent deflection} = \frac{c^2}{2R}. \qquad \textbf{(93.)} \quad \text{Art. } \textbf{1255.}$$

Or, find the chord deflection as in the preceding formula and divide it by 2. The quotient is the required tangent deflection.

STADIA MEASUREMENTS.

To find the horizontal distance between two given points, the distance between them having been read with the stadia and the vertical angle taken:

Let D = the corrected or horizontal distance;

    c = the constant;

    ak = the stadia distance;

    n = the vertical angle.

$$D = c \cos n + ak \cos^2 n. \qquad \textbf{(94.)} \quad \text{Art. } \textbf{1301.}$$

To find the difference of elevation between two given points in stadia work:

Let E = the required difference in elevation;

    c = the constant;

    ak = the stadia distance;

    n = the vertical angle.

$$E = c \sin n + ak \frac{\sin 2n}{2}. \qquad \textbf{(95.)} \quad \text{Art. } \textbf{1301.}$$

BAROMETRICAL LEVELING.

To find the difference of elevation between two points with the aneroid barometer:

Let $Z =$ the difference of elevation between the two given stations;

$h =$ the reading in inches of the barometer at the lower station;

$H =$ the reading in inches of the barometer at the higher station;

t and $t' =$ the temperature (F.) of the air at the two stations.

$$Z = (\log h - \log H) \times 60{,}384.3 \left(1 + \frac{t + t' - 64°}{900}\right).$$

(96.) Art. **1304.**

RULES AND FORMULAS USED IN SURVEYING AND MAPPING.

Rule for Balancing a Survey.—*As the sum of all the courses is to any separate course, so is the whole difference in latitude to the correction for that course. A similar proportion corrects the departures.* Art. **1315.**

Rule for Double Longitudes.—*The double longitude of the first course is equal to its departure.*

The double longitude of the second course is equal to the double longitude of the first course plus the departure of that course plus the departure of the second course.

The double longitude of the third course is equal to the double longitude of the second course plus the departure of that course plus the departure of the course itself.

The double longitude of any course is equal to the double longitude of the preceding course plus the departure of that course plus the departure of the course itself.

The double longitude of the last course (as well as of the first) is equal to its departure. This result, when obtained by the above rule, proves the accuracy of the calculation of the double longitudes of all the preceding courses. Art. **1319.**

APPLICATION OF DOUBLE LONGITUDES TO FINDING AREAS.

1. *Prepare ten columns, and in the first three write the stations, bearings, and distances, respectively.*

2. *Find the latitudes and departures of each course by the Traverse Table, placing them in the four following columns.*

3. *Balance them by the above rule for balancing a survey, correcting them in red ink.*

4. *Find the double longitudes by the rule for double longitudes, with reference to a meridian passing through the extreme east or west station, and place them in the eighth column.*

5. *Multiply the double longitude of each course by the corrected latitude for that course, placing the north products in the ninth column and the south products in the tenth column.*

6. *Add the last two columns; subtract the smaller sum from the larger, and divide the difference by 2. The quotient will be the area required.* Art. **1321.**

AREAS OF IRREGULAR FIGURES.

Trapezoidal Rule.—*Divide the figure into any sufficient number of equal parts by means of vertical lines called* **ordinates;** *add half the sum of the two end ordinates to the sum of all the other ordinates ; divide by the number of spaces (that is, by one less than the number of ordinates) to obtain the mean ordinate, and multiply the quotient by the length of the section to obtain the area.*

Simpson's Rule.—*Divide the length of the figure into any even number of equal parts, at the common distance D apart, and draw ordinates through the points of division. Add together the length of the first and the last ordinates and call the sum A ; add together the even ordinates and call the sum B ; add together the odd ordinates, except the first and the last, and call the sum C.*

Then, area of figure $= \dfrac{A + 4B + 2C}{3} \times D$. Art. **1324.**

VOLUMES OF IRREGULAR SOLIDS.

To find the volume included between two parallel cross-sections whose areas are known,

Let A = area of one section in square feet;

B = area of the other section in square feet;

C = distance between the two sections in feet;

D = required volume in cubic feet.

Then, approximately,

$$D = \frac{A + B}{2} \times C. \qquad \textbf{(97.)} \quad \text{Art. } \textbf{1325.}$$

The Prismoidal Formula.—A more accurate result than that given by the last formula is given by the prismoidal formula.

Let A = area of one section in square feet;

B = area of the other section in square feet;

M = the area of the average or mean section in square feet;

L = distance between the sections in feet;

S = the required volume in cubic feet.

$$S = \frac{L}{6}(A + 4M + B). \qquad \textbf{(98.)} \quad \text{Art. } \textbf{1326.}$$

LATITUDES AND DEPARTURES.

To find the latitude and departure of a course by means of a table of sines and cosines,

Latitude = distance × cosine bearing. **(99.)** Art. **1338.**

Departure = distance × sin bearing. **(100.)** Art. **1338.**

FORMULAS USED IN STEAM AND STEAM ENGINES.

SPECIFIC HEAT.

W = weight of body in pounds;

t = temperature before heat is applied;

t_1 = temperature after heat is applied;

c = specific heat of body;

U = number of B. T. U. required to raise temperature of body from t to t_1.

$$U = c\,W(t_1 - t). \qquad \textbf{(101.)} \quad \text{Art. } \textbf{1379.}$$

TEMPERATURE OF MIXTURES.

$w, w_1, w_2, \ldots$ = weights of the several substances, respectively;

$c, c_1, c_2, \ldots$ = specific heats of the substances, respectively;

$t, t_1, t_2, \ldots$ = temperatures of the substances, respectively;

T = final temperature of mixture.

$$T = \frac{w\,c\,t + w_1 c_1 t_1 + w_2 c_2 t_2 + \ldots}{w\,c + w_1 c_1 + w_2 c_2 + \ldots}. \quad \text{Art. } \textbf{1383.}$$

Mixture of Steam and Water.

W = weight of steam in pounds;

w = weight of water in pounds;

t_1 = temperature of steam;

t = temperature of water;

T = final temperature of mixture;

L = latent heat of steam at the given temperature.

$$T = \frac{W(L + t_1) + w\,t}{W + w}. \qquad \textbf{(102.)} \quad \text{Art. } \textbf{1384.}$$

WORK DONE BY PISTON.

p = net pressure on piston in pounds per square inch;

V = volume in cubic feet swept through by piston;

W = work done by moving piston.

$$W = 144\,p\,V. \qquad \textbf{(103.)} \quad \text{Art. } \textbf{1395.}$$

REAL AND APPARENT CUT-OFF.

s = apparent cut-off;

k = real cut-off;

i = clearance expressed as a per cent. of the stroke.

$$k = \frac{s + i}{1 + i}. \qquad \textbf{(104.)} \quad \text{Art. } \textbf{1457.}$$

HORSEPOWER.

I. H. P. = indicated horsepower of engine;

P = mean effective pressure in pounds per sq. in.;

A = area of piston in square inches;

L = length of stroke in feet;

N = number of strokes per minute.

$$\text{I. H. P.} = \frac{PLAN}{33,000}.$$ (105.)   Art. 1495.

MEAN EFFECTIVE PRESSURE.

p = gauge pressure;

k = constant depending upon cut-off (see table of Constants used in determining M. E. P.);

M. E. P. = mean effective pressure.

$$\text{M. E. P.} = .9\,[k\,(p+14.7)-17].$$ Art. 1496.

PISTON SPEED.

l = length of stroke in inches;

R = number of revolutions per minute;

S = piston speed in feet per minute.

$$S = \frac{lR}{6}.$$ (106.)   Art. 1497.

MECHANICAL EFFICIENCY OF ENGINE.

I. H. P. = indicated horsepower;

Friction H. P. = horsepower absorbed in overcoming friction of engine;

Net H. P. = I. H. P. − Friction H. P. = horsepower available to perform useful work;

E = efficiency of engine.

$$E = \frac{\text{Net H. P.}}{\text{I. H. P.}}.$$ Art. 1499.

STEAM CONSUMPTION.

l = distance between two points on the indicator diagram, one on the expansion line, and the other on the compression line, both being equally distant from the vacuum line;

L = length of indicator diagram;
a = absolute pressure of steam at the two points chosen;
W = weight of a cubic foot at pressure a;
Q = steam consumption in pounds per I. H. P. per hour.

$$Q = \frac{13{,}750 \, l \, W}{P L}. \qquad \textbf{(107.)} \quad \text{Art. } \mathbf{1507.}$$

THERMAL EFFICIENCY OF ENGINE.

T_1 = absolute temperature of steam entering cylinder;
T_2 = absolute temperature of steam leaving cylinder;
E = thermal efficiency.

$$E = \frac{T_1 - T_2}{T_1}. \qquad \text{Art. } \mathbf{1512.}$$

WATER REQUIRED BY CONDENSER.

t_1 = temperature of departing condensing water;
t_2 = temperature of entering condensing water;
t_s = temperature of the condensed steam upon leaving the condenser;
H = total heat of vaporization of one pound of steam at the pressure of the exhaust (see steam table, column 5);
W = weight of water required to condense a pound of steam.

$$W = \frac{H - t_s + 32}{t_1 - t_2}. \qquad \textbf{(108.)} \quad \text{Art. } \mathbf{1520.}$$

RATIO OF EXPANSION.

e = ratio of expansion in high-pressure cylinder;
E = total ratio of expansion;
v = volume of cylinder receiving steam from boiler;
V = volume of cylinder or cylinders exhausting into atmosphere or condenser,

$$E = \frac{e V}{v}. \qquad \textbf{(109.)} \quad \text{Art. } \mathbf{1527.}$$

FORMULAS USED IN STEAM BOILERS.

AIR REQUIRED FOR COMBUSTION AND HEAT OF COMBUSTION.

C = percentage of carbon in a fuel expressed in parts of a hundred;

H = percentage of hydrogen in a fuel expressed in parts of a hundred;

A = cubic feet of air required to burn a pound of the fuel.

$$A = 1.52\,(C + 3\,H).$$ (**110.**) Art. **1546.**

B = British thermal units produced by the combustion of the fuel;

W = weight of water that can be evaporated by a pound of the fuel.

$$B = 145\,C + 620\,H.$$ (**111.**) Art. **1547.**

$$W = \frac{B}{966}.$$ Art. **1547.**

STRENGTH OF BOILER SHELLS.

P = gauge pressure of steam, pounds per square inch;

d = diameter of shell in inches;

l = length of shell in inches;

t = thickness of material;

S = safe stress in material: 9,000 lb. for wrought iron; 11,000 lb. for steel;

F = total force tending to rupture the shell;

e = efficiency of joint (see table of Riveted Joints).

$$F = P\,d\,l.$$ (**112.**) Art. **1603.**

$$P = \frac{2\,S\,t\,e}{d}.$$ (**113.**) Art. **1604.**

HORSEPOWER OF BOILERS.

W = pounds of water evaporated per hour;

F = factor of evaporation (see table of Factors of Evaporation);

H = horsepower of boiler.

$$H = \frac{W\,F}{34.5}.$$ (**114.**) Art. **1618.**

THE SAFETY VALVE.

A = area of opening in valve-seat in square inches;

p = blow-off pressure of valve;

a = power arm of lever valve; i.e., the distance of valve-stem from fulcrum;

d = weight arm of lever valve; i. e., the distance of weight from fulcrum;

H = reading of spring scale, when the lever and valve are attached to it, at the point where the valve-stem joins the lever;

P = weight of ball hung on end of lever;

W = weight required in a dead-weight valve;

S = pounds of steam generated per hour.

$$W = pA. \qquad \textbf{(115.)} \quad \text{Art. } \textbf{1621.}$$

$$p = \frac{W}{A}. \qquad \textbf{(116.)} \quad \text{Art. } \textbf{1621.}$$

$$paA = Pd. \qquad \text{Art. } \textbf{1623.}$$

$$\left. \begin{array}{l} (pA - H)a = Pd. \\ d = \dfrac{(pA - H)a}{P}. \end{array} \right\} \qquad \textbf{(117.)} \quad \text{Art. } \textbf{1624.}$$

$$A = \frac{.5\,S}{p + 10}. \qquad \textbf{(118.)} \quad \text{Art. } \textbf{1627.}$$

DRAFT PRESSURE OF CHIMNEY.

H = height of chimney in feet;

T_a = absolute temperature of air;

T_e = absolute temperature of escaping gases;

p = draft pressure in inches of water.

$$\left. \begin{array}{l} p = H\left(\dfrac{7.6}{T_a} - \dfrac{7.9}{T_e}\right). \\[2mm] H = \dfrac{p}{\left(\dfrac{7.6}{T_a} - \dfrac{7.9}{T_e}\right)}. \end{array} \right\} \qquad \textbf{(119.)} \quad \text{Art. } \textbf{1662.}$$

QUALITY OF STEAM (BARREL CALORIMETER).

W = weight of cold water in barrel;

w = weight of mixture of steam and water run into barrel;

t = temperature of steam corresponding to observed pressure;

t_1 = original temperature of cold water;

t_2 = temperature of cold water after steam is condensed;

L = latent heat of a pound of steam at the observed pressure (see column 4, steam tables);

x = portion of weight w that is dry steam;

Q = quality of steam = $\dfrac{x}{w}$.

$$Q = \frac{x}{w} = \frac{1}{L}\left[\frac{W}{w}(t_2 - t_1) - (t - t_1)\right]. \qquad (120.)$$

Art. **1714.**

FORMULAS USED IN WATER-WHEELS.

THEORETICAL ENERGY OF A GIVEN HEAD AND WEIGHT OF WATER.

Let h = available head;

v = velocity the water would attain if it fell freely through the height h;

W = weight of water;

g = acceleration due to force of gravity = 32.16;

K = theoretical energy.

$$K = Wh = W\frac{v^2}{2g}. \qquad (121.) \quad \text{Art. } \mathbf{1727.}$$

THEORETICAL POWER.

Rule.—*To find the theoretical horsepower that a given quantity of water will furnish, multiply the weight of water that falls in one second by the distance through which it falls, and divide this product by 550; the quotient will be the theoretical horsepower.*

Let H. P. = theoretical horsepower;

Q = quantity of water falling in cubic feet per second;

H = total available fall in feet.

$$\text{H. P.} = \frac{Q \times 62.5 \times H}{550} = .1136 \, Q \, H. \quad \textbf{(122.)} \quad \text{Art.} \ \textbf{1730.}$$

ENERGY OF A JET.

Let K = energy of the jet;

W = weight of water that flows from the orifice or nozzle in one second;

w = weight of a cubic foot of water = 62.5 pounds;

a = area of the jet in square feet;

v = velocity of flow from the orifice in feet per second;

c = coefficient of velocity for the orifice;

h = head on the orifice in feet;

g = acceleration due to gravity = 32.16.

$$K = W \frac{v^2}{2g} = c \, W h. \quad \textbf{(123.)} \quad \text{Art.} \ \textbf{1731.}$$

$$W = w \, a \, v. \quad \textbf{(124.)} \quad \text{Art.} \ \textbf{1731.}$$

$$K = \frac{w \, a \, v^2}{2g} = c \, w \, a \, v \, h. \quad \textbf{(125.)} \quad \text{Art.} \ \textbf{1731.}$$

PRESSURE DUE TO IMPACT AND REACTION OF A JET.

Let P = pressure produced by the impact;

R = reaction of the jet;

W = weight of water that flows from the orifice or nozzle in one second;

w = weight of a cubic foot of water = 62.5 pounds;

a = area of the jet in square feet;

v = velocity of flow from the orifice in feet per second;

c = coefficient of velocity for the orifice;

h = head on the orifice in feet;

g = acceleration due to gravity = 32.16.

Pressure on a Vertical Surface.—When the jet impinges on a vertical surface the pressure is

$$P = w\,a\,\frac{v^2}{g} = 2\,c\,w\,a\,h = W\frac{v}{g}. \qquad \textbf{(126.)} \quad \text{Art. } \textbf{1732.}$$

Reaction.—The reaction of the jet is

$$R = P = w\,a\,\frac{v^2}{g} = 2\,c\,w\,a\,h = W\frac{v}{g}. \qquad \textbf{(127.)} \quad \text{Art. } \textbf{1732.}$$

Pressure Produced by Change of Direction.

Let $a° =$ the angle between the original direction of the jet and its direction after being deflected.

The pressure exerted on the deflecting surface in the original direction of the jet is

$$P = (1 - \cos a°)\,W\frac{v}{g}. \qquad \textbf{(128.)} \quad \text{Art. } \textbf{1734.}$$

Pressure on a Hemispherical Cup.—When the jet strikes into a hemispherical cup $a° = 180°$, and the pressure is

$$P = (1 - \cos 180°)\,W\frac{v}{g} = 2\,W\frac{v}{g};$$

that is, *the pressure is twice as great as the pressure produced when the jet struck a surface at right angles to its direction of motion.* Art. **1734.**

Effect When the Surface Is in Motion.

Let $v' =$ the velocity with which the surface moves along the line of motion of the jet.

The pressure on the surface is

$$P = (1 - \cos a°)\,W\left(1 - \frac{v'}{v}\right)\frac{v - v'}{g}. \qquad \textbf{(129.)} \quad \text{Art. } \textbf{1735.}$$

If the surface is a hemispherical cup, the pressure is

$$P = .0622\,\frac{W}{v}\,(v - v')^2. \qquad \textbf{(130.)} \quad \text{Art. } \textbf{1735.}$$

The **theoretical work** *of a jet impinging in a moving hemispherical cup is a maximum when the velocity of the cup* *is one-half the velocity of the jet, and it is equal to the theoretical work that would be done by the energy due to the velocity of the water.* Art. **1735.**

EFFICIENCY.

Rule I.—*To find the amount of work or power that can be obtained from a given fall of water when the efficiency of the motor is given, multiply the theoretical work or power by the efficiency expressed as a decimal fraction, and the product will give the available work or power.* Art. **1737.**

Rule II.—*To find the quantity of water required to furnish a given amount of power with a given efficiency, divide the theoretical quantity of water by the efficiency; the quotient will be the quantity required.* Art. **1737.**

OVERSHOT WATER-WHEELS.

Let H = total fall of water;

v = velocity of circumference of wheel;

v_e = velocity with which water enters wheel;

h = head required to produce velocity of entry v_e;

D = outside diameter of wheel;

N = number of revolutions of wheel;

Z = number of buckets;

b = breadth of buckets;

d = depth of buckets;

Q = quantity of water in cubic feet per second;

c = clearance between wheel and trough.

$$v_e = 1\tfrac{1}{2}\,v \text{ to } 2\,v. \qquad \textbf{(131.)} \quad \text{Art. } \textbf{1743.}$$

$$h = 1.1\,\frac{v_e^2}{2g}. \qquad \textbf{(132.)} \quad \text{Art. } \textbf{1743.}$$

$$D = H - (h + c). \qquad \text{Art. } \textbf{1743.}$$

$$N = 19.1\,\frac{v}{D}. \qquad \textbf{(133.)} \quad \text{Art. } \textbf{1743.}$$

$$Z = 10\,D \text{ to } 12\,D. \qquad \textbf{(134.)} \quad \text{Art. } \textbf{1743.}$$

$$d = 10 \text{ inches to } 15 \text{ inches.} \qquad \textbf{(135.)} \quad \text{Art. } \textbf{1743.}$$

$$b = 3\,\frac{Q}{dv} \text{ to } 4\,\frac{Q}{dv}. \qquad \textbf{(136.)} \quad \text{Art. } \textbf{1743.}$$

BREAST WHEELS.

The following rules may be used for the principal dimensions of a breast wheel:

Velocity of circumference of wheel $v = 2$ feet per second to 8 feet per second. Velocity of entry $v_e = 1\frac{1}{2} v$ to $2 v$.

Depth of floats $d = 10$ inches to 15 inches. Pitch of floats $t = d$.

Diameter of wheel, about twice the total head.

Breadth of wheel, $b = 1\frac{1}{2}\dfrac{Q}{dv}$ to $2\dfrac{Q}{dv}$, where Q is in cubic feet per second, b and d in feet, and v in feet per second.

Art. 1749.

UNDERSHOT WHEELS.

Let H. P. = horsepower;

v = velocity of water in race in feet per second;

v_1 = velocity of circumference of wheel in feet per second;

Q = quantity of water flowing through race in cubic feet per second;

F = area of the immersed portion of the float of a paddle wheel in an unconfined current.

For a wheel in a confined race,

$$\text{H. P.} = .00215 \, (v - v_1) \, v_1 \, Q. \qquad \textbf{(137.)} \quad \text{Art. } \textbf{1754}.$$

For a simple paddle wheel in an unconfined current,

$$\text{H. P.} = .00282 \, (v - v_1) \, v \, v_1 \, F. \qquad \textbf{(138.)} \quad \text{Art. } \textbf{1755}.$$

PONCELET'S WHEEL.

Let H = total fall in feet;

Q = the quantity of water in feet per second;

D = the outside diameter of the wheel in feet;

d = depth of floats in feet;

d_1 = depth of water current entering the wheel in feet;

v_e = the velocity of the water current entering the wheel in feet per second;

b = breadth of the wheel and of the sluice in feet;

$v_1 =$ velocity of circumference of wheel in feet per second;

$R =$ radius of curvature of floats;

$A =$ angle $A\,O\,B$ (see Fig. 527, Art. **1756**);

$u =$ number of revolutions per minute of wheel;

$n =$ number of floats in wheel.

H and Q must be determined by actual measurement; the other dimensions may then be made as follows:

$$D = 3\tfrac{1}{8}\,H;$$
$$d = \tfrac{1}{6}\,D \text{ to } \tfrac{1}{4}\,D;$$
$$d_1 = \tfrac{1}{2} \text{ foot to 1 foot};$$
$$v_e = 7.62\,\sqrt{h};$$
$$v_1 = \tfrac{1}{2}\,v_e;$$
$$b = \frac{Q}{v\,d_1};$$
$$R = 1\tfrac{1}{4}\,d;$$
$$a = 20° \text{ to } 45°;$$
$$u = 19.1\,\frac{v_1}{D};$$
$$n = 3\tfrac{1}{8}\,D.$$

Art. **1757.**

TURBINES.

In the rules and formulas used to determine the principal dimensions of reaction turbines

Let $Q =$ the available quantity of water in cubic feet per second;

$h =$ the total available head on the wheel in feet;

$v_e =$ the velocity of the flow from the guide buckets in feet per second;

$v_r =$ the relative velocity of water entering the wheel buckets in feet per second;

$v =$ the relative velocity of flow from the wheel buckets in feet per second;

$v_f =$ the absolute velocity of the water leaving the wheel buckets;

$v_w =$ the velocity of the wheel buckets at entrance;

$v'_w =$ the velocity of the wheel buckets at discharge;

a = the angle which the direction of outflow from the guides makes with the radius in a radial-flow turbine or with a perpendicular to the direction of motion of the wheel buckets in an axial-flow turbine;

a_1 = the angle which the relative direction of inflow to the wheel makes with the same lines;

a_2 = the angle which the relative direction of flow from the vanes makes with the same lines;

A = the effective outflow area of guide passages in square feet;

A_1 = the effective inflow area of wheel passages in square feet;

A_2 = the effective outflow area of wheel passages in square feet;

A_3 = sectional area of flow for draft tube in square feet;

A_4 = effective outflow area of draft tube in square feet;

N = the number of revolutions per minute;

r = the mean radius of an axial turbine in feet;

r_1 = the radius of the wheel at inflow, in feet, for a radial-flow turbine;

r_2 = the radius of the wheel at outflow, in feet, for a radial-flow turbine;

g = acceleration due to force of gravity;

K_1 = a coefficient for finding the radius r or r_1, from the area A;

k = a coefficient for finding the velocity v_e;

P = the pitch of the guide buckets;

P_1 = the pitch of the wheel buckets;

Z = the number of guide buckets;

Z_1 = the number of wheel buckets;

h_0 = the height of guide buckets in an axial-flow turbine;

h'_0 = the height of wheel buckets in an axial-flow turbine;

x = the distance between the outflow ends of two consecutive guide buckets, measured perpendicular to the direction of flow;

x_2 = the distance between the outflow ends of two consecutive wheel buckets, measured perpendicular to the direction of flow;

t = the thickness of guide buckets near ends;

t_1 = the thickness of wheel buckets near ends;

s = the part of the distance x that would be covered by the inflow end of one wheel bucket, to be measured in the same direction as x;

e = the width of outflow ends of guide buckets;

e_1 = the width of inflow ends of wheel buckets;

e_2 = the width of outflow ends of wheel buckets.

General Relations.—The usual proportions and values to be used in designing the different types of wheel are as follows:

(*a*) For axial turbines using a large quantity of water under a low head, where $\dfrac{Q}{v_e}$ is greater than 16 square feet,

$A = 70°$ to $66°$.

$A_2 = 70°$ to $66°$.

$K_1 = .67$.

$k = 1$ to $1\frac{1}{4}$.

$P = 10$ inches to 12 inches.

$t = t_1 = \begin{cases} \frac{1}{2} \text{ inch to } \frac{5}{8} \text{ inch for cast iron.} \\ \frac{1}{4} \text{ inch to } \frac{3}{8} \text{ inch for wrought iron.} \end{cases}$

$h = h_0 = \dfrac{r}{5}$ to $\dfrac{r}{6}$.

(*b*) For axial turbines using a medium quantity of water under medium head, where $\dfrac{Q}{v_e}$ is greater than 2 and less than 16 square feet,

$A = 75°$ to $70°$.

$A_2 = 74\frac{1}{2}°$ to $73°$.

$K_1 = .67$.

$k = 1.25$ to 1.5.

$P = \dfrac{r}{3.75}$ to $\dfrac{r}{4.5}$.

$t = t_1 =$ same as above.

$h = h_0 = \dfrac{r}{4}$ to $\dfrac{r}{4.5}$.

(c) For axial turbines using a small quantity of water under a high head, where $\dfrac{Q}{v_e}$ is less than 2 square feet,

$A = 75°$ to $73°$.

$A_2 = 77°$ to $74°$.

$K_1 = .67$.

$k = 1.5$ to 2.

$P = 4\frac{1}{2}$ to 6 inches.

$t = t_1 =$ same as above.

$h = h_e = \dfrac{r}{2.5}$ to $\dfrac{r}{3}$.

(d) For radial inward-flow turbines, where Q ranges from 2.4 to 275 cubic feet per second, and h is from 3 feet to 80 feet,

$A = 80°$ to $66°$.

$A_2 = 80°$ to $66°$.

$r_2 = \frac{1}{2} r_1$ to $\frac{4}{5} r_1$.

$K_1 = 0.725$ to 0.64.

$k = 0.75$ to 1.75.

$P = 4\frac{1}{2}$ inches to 12 inches.

$t = t_1 =$ same as for axial-flow turbines.

$Z_1 = Z$ to $.7 Z$.

(e) For radial outward-flow turbines, where Q ranges from 2.5 to 350 cubic feet per second, and h ranges from 3 feet to 25 feet,

$A = 75°$ to $66°$ and less.

$A_2 = 80°$ to $60°$ and less.

$r_2 = \frac{4}{3} r_1$ to $\frac{7}{8} r_1$.

$K_1 = 0.725$ to 0.64.

$k = 1.5$ to 2.

$P = \dfrac{r_1}{4.5}$ to $\dfrac{r_1}{6}$.

$t = t_1 =$ same as for axial-flow turbines.

$Z_1 = 1.2 Z$ to $1.3 Z$.   Art. **1782.**

Velocity of Entrance.—From the general relations select a value of K_1 to correspond with the type of wheel and the conditions under which it works; then,

$$v_e = K_1 \sqrt{2gh}. \qquad \textbf{(139.)} \quad \text{Art. } \textbf{1783.}$$

Effective Area.—From Q and v_e the effective area A of the passages from the guide buckets is computed from the formula

$$A = \frac{Q}{v_e}.$$ **(140.)** Art. **1783.**

Radius.—From this value of A the mean radius of a parallel-flow wheel is computed from the formula

$$r = k\sqrt{A},$$ **(141a.)** Art. **1783.**

where k is a coefficient that depends on the relation between Q and h. (See general relations.)

For a radial-flow turbine, the radius of the wheel where the water enters is given by the formula

$$r_1 = k\sqrt{A}.$$ **(141b.)** Art. **1783.**

where k depends on the style of wheel, whether outward flow or inward flow. (See general relations.)

Revolutions.—The number of revolutions per minute is given by

$$N = \frac{v_w \times 60}{2\pi r} = 9.549\frac{v_w}{r},$$ **(142a.)** Art. **1783.**

for axial-flow turbines, and for radial-flow turbines,

$$N = 9.549\frac{v_w}{r_1}.$$ **(142b.)** Art. **1783.**

Number of Vanes.—Having chosen the pitch P approximately to suit the given conditions, the number of guide vanes for an axial-flow turbine is given by the formula

$$Z = \frac{2\pi r}{P},$$ **(143a.)** Art. **1786.**

and for a radial-flow turbine the number of guide vanes is

$$Z = \frac{2\pi r_1}{P}.$$ **(143b.)** Art. **1786.**

These formulas give approximate values for Z, and the actual value is the nearest corresponding whole number.

The number of wheel vanes Z_1 for axial-flow turbines

should always be greater than Z. For ordinary cases we may take

$$Z_1 = Z + 2. \qquad (144a.) \quad \text{Art. } 1787.$$

For radial inward-flow turbines use the values

$$Z_1 = Z \text{ to } .7\,Z, \qquad (144b.) \quad \text{Art. } 1787.$$

and for radial outward-flow turbines

$$Z_1 = 1.2\,Z \text{ to } 1.3\,Z. \qquad (144c.) \quad \text{Art. } 1787.$$

Pitch.—The exact pitch for the guide vanes of axial-flow wheels is now given by

$$P = \frac{2\pi r}{Z}, \qquad (145a.) \quad \text{Art. } 1787.$$

and the pitch of wheel vanes by

$$P_1 = \frac{2\pi r}{Z_1}. \qquad (146a.) \quad \text{Art. } 1787.$$

The pitch at the outflow ends of guide vanes for radial flow turbines is

$$P = \frac{2\pi r_1}{Z}. \qquad (145b.) \quad \text{Art. } 1787.$$

For the inflow ends of the wheel vanes the pitch is

$$P_1 = \frac{2\pi r_1}{Z_1}. \qquad (146b.) \quad \text{Art. } 1787.$$

Width of Vanes.—Width of outflow end of guide vanes,

$$c = \frac{A}{Z\,x - Z_1\,s}. \qquad (147.) \quad \text{Art. } 1795.$$

The width c_1 of the inflow end of the wheel vanes is made a little greater than c, usually

$$c_1 = c + \tfrac{1}{4} \text{ inch to } c + \tfrac{3}{8}. \qquad (148.) \quad \text{Art. } 1796.$$

Width of outflow end of wheel vanes,

$$c_1 = \frac{Q}{Z_1\,x_2\,v_2}. \qquad (149.) \quad \text{Art. } 1796.$$

FORMULAS USED IN HYDRAULIC MACHINERY.

SIZE OF AIR AND VACUUM CHAMBERS.

Let V = volume of piston displacement;

V_1 = volume of air chamber;

V_2 = volume of vacuum chamber.

For ordinary double-acting pumps working under moderate pressures at ordinary speeds,

$$V_1 = 3\,V. \qquad \text{Art. } \mathbf{1885.}$$

For pressures of 100 pounds per square inch and upwards, or for high piston speeds,

$$V_1 = 6\,V. \qquad \text{Art. } \mathbf{1885.}$$

For ordinary cases, make

$$V_2 = \tfrac{1}{2}\,V_1. \qquad \text{Art. } \mathbf{1889.}$$

CALCULATIONS RELATING TO PUMPS.

Displacement.

Let D = displacement in cubic feet per minute;

d = diameter of piston or plunger in inches;

L = length of stroke of piston or plunger in inches;

N = number of discharge strokes made by piston or plunger in 1 minute.

Then, $D = .000455\,d^2\,L\,N.$ Art. **1905.**

Slip.

Let s = slip;

D = displacement;

C = actual discharge.

Then, $\qquad s = \dfrac{D - C}{D}.$ Art. **1909.**

Head and Pressure.

Let H = head in feet;

P = pressure in pounds per square inch.

$$P = .434\,H. \qquad \text{Art. } \mathbf{1914.}$$
$$H = 2.304\,P. \qquad \text{Art. } \mathbf{1915.}$$

Size of Piston or Plunger.

Let G = number of gallons discharged per minute;

S = speed in feet per minute of piston or plunger;

d = diameter of piston or plunger in inches;

F = number of cubic feet discharged per minute.

Then, the theoretical diameter of piston or plunger is

$$d = 4.95 \sqrt{\frac{G}{S}} = 13.54 \sqrt{\frac{F}{S}}. \qquad \text{Art. } \mathbf{1916.}$$

If we add 25 per cent. to the required discharge to allow for slip, the diameter of the piston or plunger will be

$$d = 5.535 \sqrt{\frac{G}{S}} = 15.138 \sqrt{\frac{F}{S}}. \qquad \textbf{(152.)} \quad \text{Art. } \mathbf{1916.}$$

Discharge.

The theoretical discharge in cubic feet equals the displacement.

The theoretical discharge in gallons per minute is

$$G = .0408 \, d^2 S. \qquad \text{Art. } \mathbf{1917.}$$

If we make the same allowance for slip as was made in formula **152**, the discharge in gallons per minute is

$$G = .03264 \, d^2 S; \qquad \textbf{(153.)} \quad \text{Art. } \mathbf{1917.}$$

and in cubic feet per minute,

$$F = .00436 \, d^2 S. \qquad \text{Art. } \mathbf{1917.}$$

Power.

Let H. P. = horsepower;

H = vertical height in feet from the surface of the water in the well or sump to the center of the outlet end of discharge pipe;

G = discharge in gallons per minute;

F = discharge in cubic feet per minute.

The theoretical power is

$$\text{H. P. } = .000254 \, G H = .0019 \, F H. \qquad \text{Art. } \mathbf{1918.}$$

If, for ordinary cases, the frictional resistances are

assumed to be 50 per cent. of the power developed by the engine, the power required is

$$\text{H. P.} = .00038 \, G \, H. \qquad (154.) \quad \text{Art. } 1918.$$

To find the height to which a given power will raise a given quantity of water, making the same allowance for friction as in the last formula,

$$H = \frac{\text{H. P.}}{.00038 \, G}. \qquad (155.) \quad \text{Art. } 1919.$$

Size of Steam Cylinder.

Let S = steam piston speed;

d = diameter of steam cylinder in inches;

r = ratio between length of stroke and diameter of cylinder;

l = length of stroke in feet;

N = number of strokes per minute;

H. P. = horsepower;

P = steam pressure in pounds per square inch.

Then, for simple direct-acting steam pumps,

$$d = \sqrt[3]{\frac{504,201.6 \times \text{H. P.}}{r \, P \, N}}; \qquad (156.) \quad \text{Art. } 1920.$$

$$\text{or, } d = \sqrt{\frac{42,016.8 \times \text{H. P.}}{P \, S}}. \qquad (157.) \quad \text{Art. } 1920.$$

Having obtained the diameter of the steam piston by either of the above formulas, the stroke can be found by multiplying the diameter by the value of the ratio r. When formula **157** is used, the number of strokes can be found by dividing the piston speed by the length of the stroke in feet.

Sizes of Suction and Delivery Pipes.

For a velocity of 200 feet per minute in the suction pipe and 400 feet per minute in the delivery pipe,

Let d_1 = diameter of suction pipe;

d_2 = diameter of delivery pipe;

G = discharge in gallons per minute.

$$d_1 = 4.95 \sqrt{\frac{G}{200}}, \text{ or } d_1 = .35 \sqrt{G}. \qquad (158.) \quad \text{Art. } 1921.$$

$$d_2 = 4.95 \sqrt{\frac{G}{400}}, \text{ or } d_2 = .25 \sqrt{G}. \qquad (159.) \quad \text{Art. } 1921.$$

The pipes may be made larger than the values calculated by the above formulas, particularly the suction pipe, but it is not good practice to make them any smaller.

DUTY OF A PUMP.

Old Standard.—According to the old standard, the duty of a pumping engine is *the number of pounds of water raised one foot high for each 100 pounds of coal burned in the boiler.*

Let G = number of gallons discharged in a given period·

h = total vertical distance in feet from the surface of the water in the well, or other source of supply, to the point of discharge;

W = the number of pounds of coal burned in the given period;

D = the duty in foot-pounds.

Then,

$$D = \frac{835.5\,G\,h}{W}. \qquad (160.) \quad \text{Art. } 1922.$$

Standard Recommended by Committee of American Society of Mechanical Engineers.

The duty of a pumping engine is equal to the total number of foot-pounds of work actually done by the pump, divided by the total number of heat units in the steam used by the pump, including the steam used by the condensers (if any) and boiler feed, and this quotient multiplied by 1,000,000.

The number of foot-pounds of work done by the pump is to be found as follows: A pressure gauge is attached to the discharge pipe and a vacuum gauge to the suction pipe, both as near the pump as convenient; then the pressure against which the pump plunger works is equal to the difference in the pressures shown by these two gauges plus the head due to the difference in level of the points in the pipes to which they are attached; and the number of foot-pounds is equal to the continued product of the net area of the plunger (making allowance for piston rods), the length of the plunger stroke in feet, the number of plunger strokes made during the trial, and the pressure against which the pump plungers work, as shown by the gauges.

The number of heat units furnished to the pump is the number of British thermal units (B. T. U.) in the steam from the boilers, and is to be determined by an evaporation test of the boilers. If we let

A = the net area of the plunger in square inches;

P = the pressure in pounds per square inch indicated by the gauge on the discharge pipe;

p = the pressure in pounds per square inch corresponding to the vacuum indicated by the gauge on the suction pipe;

S = the pressure in pounds per square inch corresponding to the difference in level between the two gauges;

L = the average length of stroke of pump plunger in feet;

N = the total number of single strokes of plunger made during the trial;

H = the total number of heat units consumed by the engine during the trial;

W = the total number of foot-pounds of work done by the pump during the trial; and

D = the duty.

Then, $W = A (P \pm p + S) L N$,   **(161.)**   Art. **1924.**

and $$D = \frac{W}{H} \times 1,000,000 =$$

$$\frac{A (P \pm p + S) L N}{H} \times 1,000,000. \qquad \textbf{(162.)} \quad \text{Art. } \textbf{1924.}$$

CALCULATIONS RELATING TO HYDRAULIC MACHINERY.

Relations Between Pressure and Size of Ram.

Let D = the diameter of a hydraulic piston or ram;

W = the weight of the ram and attachments that must be lifted by the water;

$p =$ the pressure of the water in pounds per square inch;

$F =$ the percentage of friction;

$P =$ the net pressure exerted by the ram.

To find the net pressure exerted by a ram or plunger of a hydraulic press,

$$P = .7854 \times D^2 \times p \times \left(1 - \frac{F}{100}\right) - W. \qquad \textbf{(163.)}$$
$$\text{Art. } \textbf{1969.}$$

To find the pressure per square inch required to exert a given net pressure when the diameter and weight of the ram and the percentage of friction are given, use the formula

$$p = \frac{P + W}{.7854\, D^2 \times \left(1 - \frac{F}{100}\right)}. \qquad \textbf{(164.)} \quad \text{Art. } \textbf{1970.}$$

To find the diameter of piston or ram required to exert a given net pressure, use the formula

$$D = \sqrt{\frac{P + W}{.7854 \times p \times \left(1 - \frac{F}{100}\right)}}. \qquad \textbf{(165.)} \quad \text{Art. } \textbf{1971.}$$

Weight and Volume of Accumulators.

Let $W_1 =$ weight of accumulator ram;

$W_2 =$ load on accumulator ram;

$D_1 =$ diameter of accumulator ram;

$p_1 =$ maximum pressure per square inch in the accumulator cylinder;

$p_2 =$ minimum pressure per square inch in the accumulator cylinder;

$p =$ mean pressure per square inch in the accumulator cylinder;

$S =$ stroke of accumulator ram;

$V =$ total volume of water displaced by accumulator ram during the stroke S;

$F =$ the percentage of friction.

To find the mean pressure p corresponding to a given case, use the formula

$$p = \frac{W_1 + W_2}{.7854\, D_1^2}. \qquad \textbf{(166.)} \quad \text{Art. } \textbf{1973.}$$

The maximum pressure is found by the formula

$$p_1 = \frac{W_1 + W_2}{.7854\, D_1^2 \times \left(1 - \frac{F}{100}\right)}, \qquad \textbf{(167.)} \quad \text{Art. } \textbf{1973.}$$

and the minimum pressure by

$$p_2 = \frac{W_1 + W_2}{.7854\, D_1^2 \times \left(1 + \frac{F}{100}\right)}. \qquad \textbf{(168.)} \quad \text{Art. } \textbf{1973.}$$

The **weight** required to produce a given mean pressure when diameter and weight of the ram are known may be found from the following formula:

$$W_2 = .7854 \times D_1^2 \times p - W_1. \qquad \textbf{(169.)} \quad \text{Art. } \textbf{1974.}$$

The relations between the stroke, diameter, and volume of an accumulator are given by the following formulas:

$$V = .7854\, D_1^2\, S. \qquad \textbf{(170.)} \quad \text{Art. } \textbf{1977.}$$

$$S = \frac{V}{.7854\, D_1^2}. \qquad \textbf{(171.)} \quad \text{Art. } \textbf{1977.}$$

$$D_1 = 1.128 \sqrt{\frac{V}{S}}. \qquad \textbf{(172.)} \quad \text{Art. } \textbf{1977.}$$

In the above formulas, if D_1 and S are in inches, the volume will be given in cubic inches; and if D_1 and S are in feet, V will be given in cubic feet.

FORMULAS USED IN WATER SUPPLY AND DISTRIBUTION.

DIMENSIONS OF SPILLWAY OR OVERFLOW.

Let L = length of lip of spillway in feet;

A = area of watershed above dam in square miles;

D = depth of notch of spillway in feet;

Q = cubic feet of water per second per square mile;

C = a constant depending on the character of the dam and its surroundings and the area of the watershed.

Then, $L = 20\sqrt{A}.$ (173.)   Art. **2048.**

$$D = \frac{\sqrt[3]{Q^2 \times A}}{16} + C.$$ (174.)   Art. **2048.**

If we assume $Q = 64$, which corresponds to a little over 41 million gallons per 24 hours, per square mile, and represents a very powerful freshet flow, although, perhaps, not the maximum, formula **174** reduces to

$$D = \sqrt[3]{A} + C.$$ (175.)   Art. **2049.**

MASONRY DAMS.

Let A = thickness of top of a trapezoidal dam in feet;
  B = thickness of base of dam in feet;
  C = a factor of safety against either sliding or overturning;
  D = density (weight per cubic foot) of material of which dam is built;
  H = head of water pressing against the dam in feet;
  R = resistance of wall to sliding;
  T = horizontal thrust in pounds on the dam, due to the head H;
  MR = moment of resistance of dam against overturning by rotating about its outer toe;
  MT = moment of thrust about the outer toe of the dam.

The thrust is

$$T = 31.25\,H^2,$$ (176.)   Art. **2063.**

and the moment of thrust

$$MT = 10.42\,H^3.$$ (177.)   Art. **2063.**

The resistance of the wall to sliding is

$$R = 0.75\,A\,D.$$ (178.)   Art. **2065.**

The moment of resistance to overturning for a wall with vertical sides is

$$MR = \frac{D\,H\,B^2}{2},$$ (179.)   Art. **2066.**

and for a trapezoidal wall

$$MR = \frac{DH}{3}\left(AB - \frac{A^2}{2} + B^2\right).$$ **(180.)**   Art. **2066.**

The relation between A, B, D, and H for a factor of safety C against sliding is given by the formula

$$B = \frac{83.33\,CH}{D} - A.$$ **(181.)**   Art. **2067.**

For a factor of safety C against overturning, the breadth of the base is given by the formula

$$B = \tfrac{1}{2}\sqrt{\frac{125\,CH^2}{D} + 3A^2} - \frac{A}{2}.$$ **(182.)**   Art. **2068.**

Average Dimensions.—For practical values of A and D, a satisfactory value of B is

$$B = \tfrac{2}{3}H \text{ to } \tfrac{3}{4}H.$$ Art. **2070.**

HIGH MASONRY DAMS.

Maximum Unit Stress on Base of Dam for Unequally Distributed Load.

Let L = length of base of a section through the dam;

     d = length of the shorter segment of this base;

$L-d$ = length of the remaining segment;

     W = the resultant of the weight of the section, or the vertical component of this resultant;

     P = maximum unit stress.

There are three empirical formulas for the value of P, which experience shows give satisfactory results; viz.:

$$P = \frac{4\,W}{L^2}(L - 1.5\,d),$$ **(183.)**   Art. **2072.**

$$P = \frac{2\,W}{3\,d},$$ **(184.)**   Art. **2072.**

and $$P = \frac{W(L - d)}{Ld}.$$ **(185.)**   Art. **2072.**

Of these formulas, the last is probably the most satisfactory.

DARCY'S FORMULAS.

Let D = diameter of pipe in feet;

　H = total head in feet;

　L = total length in feet;

　V = velocity of efflux in feet per second;

　C = an experimental coefficient (see table of Coefficients for Darcy's Formula);

　Q = quantity discharged in cubic feet per second;

　A = area of pipe in square feet;

　h = head per 1,000 feet of length $= \dfrac{H}{1,000\,L}$.

Fundamental Formulas for Long Pipes.

$$\frac{D H}{C L V^2} = 1. \qquad \textbf{(186.)} \quad \text{Art. } \mathbf{2092.}$$

$$V = \sqrt{\frac{D H}{C L}}. \qquad \textbf{(187.)} \quad \text{Art. } \mathbf{2092.}$$

$$Q = A \sqrt{\frac{D H}{C L}}. \qquad \textbf{(188.)} \quad \text{Art. } \mathbf{2092.}$$

$$Q = 0.7854\, D^2 \sqrt{\frac{D H}{C L}}. \qquad \textbf{(189.)} \quad \text{Art. } \mathbf{2092.}$$

$$Q = \sqrt{\frac{.617\, D^5 H}{C L}}. \qquad \textbf{(190.)} \quad \text{Art. } \mathbf{2092.}$$

Approximate Formulas for Rough Pipes

For pipes from 8 inches to 48 inches in diameter,

$$Q = \sqrt{\frac{1,000\, D^5 H}{L}}. \qquad \textbf{(191.)} \quad \text{Art. } \mathbf{2094.}$$

$$Q = \sqrt{D^5 h}. \qquad \textbf{(192.)} \quad \text{Art. } \mathbf{2094.}$$

$$\frac{Q^2}{h\, D^5} = 1. \qquad \textbf{(193.)} \quad \text{Art. } \mathbf{2094.}$$

$$Q = D^2 \sqrt{D h}. \qquad \textbf{(194.)} \quad \text{Art. } \mathbf{2094.}$$

$$\frac{Q}{D^2 \sqrt{D h}} = 1. \qquad \textbf{(195.)} \quad \text{Art. } \mathbf{2094.}$$

For pipes from 3 inches to 6 inches in diameter,

$$\frac{Q^2}{hD^5} = 0.785.$$ (**196.**) Art. **2094.**

$$Q = 0.89\sqrt{D^5h}.$$ (**197.**) Art. **2094.**

Formulas for Smooth Pipes.

$$Q = \sqrt{2\,D^5h}.$$ (**198.**) Art. **2095.**

$$\frac{Q^2}{hD^5} = 2.$$ (**199.**) Art. **2095.**

$$Q = 1.40\sqrt{D^5h}.$$ (**200.**) Art. **2095.**

General Relation Between Smooth and Rough Pipes.

In general, the discharge through a smooth pipe is 1.40 times that through a rough pipe of the same diameter; and, reciprocally, the discharge through a rough pipe is 0.70 times that through a smooth one of the same diameter. These factors represent the practical limits between which the extremes of smoothness and roughness can affect the flow. Art. **2095.**

Formulas for Velocity.

For rough pipes of from 8 inches to 48 inches in diameter,

$$V = 1.27\sqrt{Dh}.$$ (**201.**) Art. **2096.**

For rough pipes of smaller diameter,

$$V = 1.13\sqrt{Dh}.$$ (**202.**) Art. **2096.**

For smooth pipes of large diameter,

$$V = 1.78\sqrt{Dh}.$$ (**203.**) Art. **2096.**

For smooth pipes of small diameter,

$$V = 1.60\sqrt{Dh}.$$ (**204.**) Art. **2096.**

General Relation Between the Elements of Two Pipes.

Let D, Q, L, H, and C be the respective elements of one pipe and D', Q', L', H', and C' the similar elements of another; then,

$$\frac{DHC'L'V'^2}{D'H'CLV^2} = 1.$$

If, as can usually be done, we make $C = C'$, we have

$$\frac{DHL'V'^2}{D'H'LV^2} = 1. \qquad \textbf{(205.)} \quad \text{Art. } \textbf{2097.}$$

Also, $\dfrac{Q^2LD'^4H'}{Q'^2L'D^5H} = 1. \qquad \textbf{(206.)} \quad \text{Art. } \textbf{2097.}$

If L and H equal, respectively, L' and H',

$$\frac{Q'}{Q} = \sqrt{\frac{D'^5}{D^5}}. \qquad \textbf{(207.)} \quad \text{Art. } \textbf{2098.}$$

To find the number x of small pipes with the diameter D' to replace a pipe whose diameter is D,

$$x = \sqrt{\frac{D^5}{D'^5}}. \qquad \textbf{(208.)} \quad \text{Art. } \textbf{2098.}$$

COMPOUND PIPES.

To find the diameter of a simple pipe that will give the same delivery as a given compound system:

Let D = diameter of the simple pipe;

L = length of the simple pipe;

d, d', d'', etc. = diameters of the respective sections of the compound pipe;

l, l', l'', etc. = lengths of the respective sections of the compound pipe.

Then,

$$\frac{L}{D^5} = \frac{l}{d^5} + \frac{l'}{d'^5} + \frac{l''}{d''^5} + \text{etc.} \qquad \textbf{(209.)} \quad \text{Art. } \textbf{2111.}$$

PUMPING INTO MAINS.

Theoretical horsepower required to force a given quantity of water into a main against a given pressure head:

Let H. P. = theoretical horsepower;

H = pressure head in feet;

Q = quantity of water in cubic feet per second.

$$\text{H. P.} = \frac{QH}{8.8}. \qquad (210.) \quad \text{Art. } 2117.$$

WEIGHTS AND THICKNESS OF CAST-IRON PIPES.

Let W = weight in pounds;

D = diameter in inches;

T = thickness in inches;

L = length in inches;

P = weight in long tons (2,240 pounds);

M = length in miles;

W' = approximate weight per foot in pounds;

H = total head in feet.

$$W = 0.82 (D + T) T \times L. \qquad (211.) \quad \text{Art. } 2125.$$

$$W' = 10 (D + T) T. \qquad (212.) \quad \text{Art. } 2126.$$

$$P = 25 M (D + T) T. \qquad (213.) \quad \text{Art. } 2127.$$

$$T = 0.00006 HD + 0.0133 D + 0.296. \qquad (214.) \quad \text{Art. } 2128.$$

DARCY'S FORMULAS FOR FLOW IN OPEN CHANNELS.

Let U = mean velocity of flow in feet per second;

S = water section in square feet;

WP = wet perimeter in feet;

R = mean hydraulic radius $= \dfrac{S}{WP}$;

I = slope of free water surface per foot of length = total fall of surface divided by total length;

D = interior diameter of a circular conduit in feet.

For an ordinary tunnel or channel lined with well-laid brick,

$$U = R \sqrt{\frac{100,000 \, I}{6.6 \, R + 0.46}}.$$ **(215.)** Art. **2143.**

For a circular brick-lined conduit running full,

$$U = \frac{D}{4} \sqrt{\frac{100,000 \, I}{1.65 \, D + 0.46}}.$$ **(216.)** Art. **2144.**

FORMULAS USED IN IRRIGATION.

APPROXIMATE DISCHARGE OF WEIRS.

Let l = length of notch in feet;

H = measured head on crest in feet;

Q = discharge in cubic feet per second;

then, $Q = 3\frac{1}{3} \, l \, H^{\frac{3}{2}}$. **(217.)** Art. **2163.**

FLOW OF WATER THROUGH CONDUITS.

Let h = difference in level between the ends of the canal, or any two cross-sections of the canal;

l = horizontal length of that portion of the canal included between the sections whose difference of level is h;

s = slope = the ratio $\dfrac{h}{l}$;

a = area of the water cross-section;

p = wetted perimeter;

r = hydraulic radius = the ratio $\dfrac{a}{p}$;

c' = a coefficient depending on the nature of the surface of the conduit;

and v = mean velocity of flow.

The laws for the resistance to flow may be expressed by the relation $h \, a = c' \, l \, p \, v^2$, from which we have the general formula

$$v = \sqrt{\frac{h}{c'l} \times \frac{a}{p}} = \sqrt{\frac{1}{c'} \times s \times r}.$$ **(218.)** Art. **2173.**

By replacing $\sqrt{\dfrac{1}{c^2}}$ by the equivalent factor c, we have $v = c\sqrt{rs}$, the same as formula **50**, Art. **1033**.

Formulas for Flow in Canals.

Canals with earthen banks,

$$v = \sqrt{\frac{100,000\,r^2 s}{9\,r + 35}}. \qquad \textbf{(219.)} \quad \text{Art. } \textbf{2177.}$$

Canals lined with dry stone,

$$v = \sqrt{\frac{100,000\,r^2 s}{8\,r + 15}}. \qquad \textbf{(220.)} \quad \text{Art. } \textbf{2183.}$$

Canals lined with rubble masonry,

$$v = \sqrt{\frac{100,000\,r^2 s}{7.3\,r + 6}}. \qquad \textbf{(221.)} \quad \text{Art. } \textbf{2184.}$$

Wooden flumes,

$$v = \sqrt{\frac{100,000\,r^2 s}{6.6\,r + 0.46}}, \qquad \textbf{(226.)} \quad \text{Art. } \textbf{2196.}$$

TIMBER FOR FLUMES.

Let W = total load in pounds carried by any beam;
  l = length of beam in inches;
  b = breadth of beam in inches;
  d = depth of beam in inches;
  S = maximum unit fiber stress in pounds per square inch.

For a simple beam with a uniformly distributed load,

$$W = \tfrac{4}{3}\frac{b\,d^2}{l}S. \qquad \textbf{(222.)} \quad \text{Art. } \textbf{2189.}$$

For a simple beam with a concentrated load at the middle,

$$W = \tfrac{2}{3}\frac{b\,d^2}{l}S. \qquad \textbf{(223.)} \quad \text{Art. } \textbf{2190.}$$

For a beam with a concentrated load at a distance l_1 from one support and l_2 from the other, where $l_1 + l_2 = l$,

$$W = \tfrac{1}{8}\frac{b\,d^2\,l\,S}{l_1\,l_2}. \qquad \textbf{(224.)} \quad \text{Art. } \textbf{2191.}$$

For a beam on which the load at one end is zero, with a uniform increase in the load to the other end,

$$W = 1.3\,\frac{b\,d^2}{l}\,S. \qquad \textbf{(225.)} \quad \text{Art. } \textbf{2192.}$$

SAFE WORKING STRESS S.

For good sound timber.

Kind of Timber.	Safe Working Stress.	
	Steady Load.	Variable Load.
Yellow Pine..................	1,800	1,200
White Oak..................	1,350	1,000
Spruce	1,250	900
Hemlock	1,200	850
White Pine	1,100	800

TRUSSES.

Trussed Stringers.

Let L = span in inches;

H = depth of truss in inches;

b = breadth of stringer in inches;

d = depth of stringer in inches;

W_t = total uniformly distributed load in pounds;

S = allowable unit fiber stress in stringer or strut;

S_2 = total stress in tie-rods;

h = width of strut in inches;

t = thickness of strut in inches.

Relation between W_t and dimensions of stringer,

$$W_t = \frac{32\,b\,d^2\,H}{3\,L\,(2\,H + d)}\,S. \qquad \textbf{(227.)} \quad \text{Art. } \textbf{2198.}$$

Stress in tie-rods,

$$S_2 = \tfrac{5}{16}\frac{W_t}{H}\sqrt{\left(\frac{L}{2}\right)^2 + H^2}. \qquad \textbf{(228.)} \quad \text{Art. } \textbf{2198.}$$

Stress in strut,

$$W_t = \tfrac{8}{3}\,h\,t\,S. \qquad \textbf{(229.)} \quad \text{Art. } \textbf{2198.}$$

The King-Rod Truss.

Let L = length of span in inches;

$\quad W_t$ = total uniformly distributed load in pounds;

$\quad W$ = total stress in each strut in pounds;

$\quad S_s$ = safe unit stress in king-rods;

$\quad A$ = net sectional area of king-rods;

and $\quad H$ = depth of truss in inches;

then, $\quad W = \tfrac{5}{16}\frac{W_t}{H}\sqrt{\left(\frac{L}{2}\right)^2 + H^2}, \qquad \textbf{(230.)} \quad \text{Art. } \textbf{2199.}$

and $\qquad A = \tfrac{5}{8}\frac{W_t}{S_s}. \qquad \textbf{(231.)} \quad \text{Art. } \textbf{2199.}$

The Queen-Rod Truss.

Let S_t = maximum unit stress in tie-beam;

$\quad W$ = total uniformly distributed load in pounds;

$\quad L$ = length of span in inches;

$\quad H$ = depth of span in inches;

$\quad b$ = breadth of tie-beam in inches;

$\quad d$ = depth of tie-beam in inches;

$\quad S_c$ = total stress in upper chord member in pounds;

$\quad S_s$ = total stress in struts in pounds;

$\quad S_q$ = allowable unit stress in queen-rods;

$\quad A$ = sectional area of queen-rods.

The maximum unit stress in tie-beam is

$$S_t = \tfrac{1}{4}\frac{WL}{b\,d}\left(\frac{1}{2H} + \frac{1}{3d}\right). \qquad \textbf{(232.)} \quad \text{Art. } \textbf{2200.}$$

The total stress in the upper chord member is

$$S_c = \tfrac{1}{8}\frac{WL}{H}. \qquad \textbf{(233.)} \quad \text{Art. } \textbf{2200.}$$

The total stress in each strut is

$$S_s = W'\sqrt{\tfrac{1}{9} + \tfrac{1}{64}\frac{L^2}{H^2}}. \qquad \textbf{(234.)} \quad \text{Art. } \textbf{2200.}$$

The net area of each queen-rod is

$$A = \tfrac{1}{3}\frac{W}{S_q}. \qquad \textbf{(235.)} \quad \text{Art. } \textbf{2200.}$$

The Howe Truss.

Let N_t = the number of a tie, counting from the center;
  S_t = total stress in a tie;
  L_t = length of a tie ;
  L_s = length of a strut;
  S_s = total stress in a strut;
  N = number of panels in truss from center to either abutment;
  n = number of panels from a given panel to the nearer abutment ;
  P = panel load in pounds ;
  L_p = length of a panel ;
  S_{tc} = total stress in top chord of a given panel ;
  S_{bc} = total stress in bottom chord of a given panel.

The stress in any tie whose number is N_t is

$$S_t = (N_t + \tfrac{1}{2})\,P. \qquad \textbf{(236.)} \quad \text{Art. } \textbf{2202.}$$

The stress in a strut is

$$S_s = \frac{L_s}{L_t}S_t. \qquad \textbf{(237.)} \quad \text{Art. } \textbf{2202.}$$

The top chord stress is

$$S_{tc} = n\,P\,(N - \tfrac{1}{2}n)\frac{L_p}{L_t}. \qquad \textbf{(238.)} \quad \text{Art. } \textbf{2202.}$$

The bottom chord stress is

$$S_{bc} = P(Nn + N - n - \tfrac{1}{2}n^2 - \tfrac{1}{2})\frac{L_p}{L_t}. \qquad (239.) \quad \text{Art. } 2202.$$

REFUSAL OF PILES.

Let S = weight a pile will bear with safety;

W = weight of hammer, in the same unit as S;

H = height of fall of hammer in feet;

then, $S = WH.$ (240.)   Art. 2205.

INDEX.

www.ingramcontent.com/pod-product-compliance
Lightning Source LLC
Chambersburg PA
CBHW021811190326
41518CB00007B/549